Hans Sommer – Project Management for Building Construction

Hans Sommer

Project Management for Building Construction

35 Years of Innovation at Drees & Sommer

Springer

Prof. Dr. -Ing. Hans Sommer
Drees & Sommer AG
Untere Waldplätze 28
70569 Stuttgart
Germany
hans.sommer_V@dreso.com

ISBN 978-3-642-10873-0   e-ISBN 978-3-642-10874-7
DOI 10.1007/978-3-642-10874-7
Springer Heidelberg Dordrecht London New York

Library of Congress Control Number: 2009942360

Cover design: Drees & Sommer AG, Stuttgart

Printed on acid-free paper

Springer is part of Springer Science+Business Media (www.springer.com)

Preface

Construction has turned into an ever more complex mesh of relationships between increasingly accelerating processes, decisions and actions. At the same time, however, there is a development toward sustainable design that leads to buildings providing the best possible connection of functionality and architecture, energy efficiency and healthy construction materials that can be recycled while at the same time also achieving the best possible economical benefits.

Following its modest beginnings, the Drees & Sommer corporation has grown in this area and for over 35 years now has been significantly contributing to the development of modern project management while always putting an emphasis on innovation when it came to management method. On this basis, the third edition of this book has been completely reviewed and now constitutes the current status of development.

The key to successfully handling current complexities and dynamics is with the project managers who are completely at home in the subject of contents and inter-relations between planning and construction implementation and who master state-of-the-art management methods. If, in addition to this, constructive cooperation between all participants in the team can be achieved, there are the best possible chances of achieving the optimum possible project result.

At major structural engineering projects, project participants from the most different areas of interests and knowledge gather in one place: Architects, project managers and specialized planners, representatives of the client, of the relevant authorities and also from the building and construction industry. Communication difficulties cannot be ruled out in such a heterogeneous circle. It is, hence, one of first aims of this book, to outline both the participants and the process of structural engineering projects – for planning and construction – by using striking examples to describe them clearly. Furthermore, the essential management tasks and possible management variants are explained.

With this book, I would like to thank all employees and partners of the Drees & Sommer group who have accompanied me over the course of this development and have fundamentally contributed to the contents and professional implementation of the same.

Furthermore I would like to thank Ms Angela Reitmaier, for the preparation of the layout and the illustrations, to the stage of galley proofs ready for print, as well as Steffen Sommer for his editing of the texts.

Stuttgart, June 2009

Success by Innovation

With Drees & Sommer, the author, with numerous innovations, was directionally involved for over 35 years in many areas of the evolution of today's "project management".

The response of the property market to the project management idea was anything but enthusiastic at the beginning of the seventies. Of course, this was primarily also because young civil engineers tried suddenly without too much construction experience, to decode the structure of the building processes and to prepare exact schedules by means of the network planning method that had arisen from the space-travel industry. The constant feedback that is required from the site supervisors and site foremen took up their time without their being able to recognize an actual benefit, as opposed to control by handshake that had been in place thus far.

The increasing automation of buildings at the same time, with air conditioners, curtain facades, – modern lighting, ceiling and partition wall systems – soon resulted in structured time scheduling and control on the construction site so that a new consultation area opened up by around 1974. We soon recognized, however, that the control of the construction sites could only work if planning were also integrated into the process structures. Architects and engineers were suddenly affected, too. There were plan delivery lists and contract appointments; From 1975, we called our service "project control" and organized perfect processes.

In the media at that time, considerable construction cost increases were discussed for a number of prominent public projects. We focused on the problem and recognized that the construction costs were being completely insufficiently assessed by the architects via cubic meter of enclosed space although the complexity of the projects could not be represented by this any longer. Cost esti-

mates and records of detailed measures were worlds apart. From 1977, we were among the protagonists of cost assessment for building elements, for which we developed a software of our own as of 1979. The so-called element method was taken over into the DIN 276, in 1981, by the standards committee: costs of rising structures. In 1983, in his thesis "cost control of rising structures", the author developed a uniform method, from cost assessment according to elements up to calls for tender for the major positions and trades which, today, is still in use.

Parallel to this development, we started at the beginning of the eighties to focus on building operation costs and later on building use costs and – with profitability consulting and facility management consulting – to establish new service areas on the market. We had recognized that the subsequent costs add up to enormous sums over the complete lifetime of a building if no corresponding measures, which we nowadays describe as "life cycle engineering", are taken up. The reduction of energy costs was influenced quite substantially by simulation software developed at Drees & Sommer for energy optimization, which was already used as of 1986.

Finally, we developed the first project communication systems PIS and PCM, without which project execution today is hardly conceivable any more, and catapulted CAD video visualization into new dimensions with projects like Potsdamer Platz in Berlin.

At first, the innovation dynamics of Drees & Sommer were driven by some individual creative partners before being put on a broader basis from 1999, with the help of our award-winning knowledge management and the worldwide network of employees.

35 Years Drees & Sommer-Innovations in Project Management

1974 ——— Innovation – Calculation of network plans on industrial computers

1975 ——— Service: "Project Control"

1976 ——— Service: "Cost Planning"

1977 ——— Innovation – Cost assessment with cost elements

1978 ——— Innovation – Software KOPLAN for cost assessment

1979 ——— Innovation – Application of mean data technology PDP 11

1980 ——— Innovation – Software TEPLAN for schedule planning

1981 ——— Service: "Building Use Costs"

1982 ——— Innovation – Software BEKOS for operating cost assessment

1983 ——— Innovation – Profitability calculation

1986 ——— Innovation – Energetic building simulation

1988 ——— Service: "Facility Management Consulting"

1993 ——— Service: "Development Management"

1994 ——— Innovation – CAD-visualizations

1995 ——— Service: "General Construction Management"

1996 ——— Innovation – Project information system on Hypercard

1997 ——— Service: "Optimization User Requirements"

1999 ——— Innovation – Award-winning knowledge management

1999 ——— Service: "Energy-Saving Construction"

2000 ——— Innovation – Award-winning PCM

2003 ——— Service: "Risk Management"

2003 ——— Innovation – Software for Life Cycle Costs

2004 ——— Innovation – Cost Monitor for professional cost control

2006 ——— Innovation – KAIZEN for effective process organization of construction projects

2007 ——— Service: "Green Building Consulting and International Certification"

2008 ——— Innovation – Interactive project manual for project management

2009 ——— Innovation – E-Learning for all competences with Drees & Sommer

Content

1 Fundamentals of Project Execution

In order to assess the topic of project management better, one initially needs to have a certain idea of the general problematic and environment. What exactly can a client expect when he or she wishes to undertake a property project? What are the tasks he/she is responsible for as a client, so that he/she really gets what he/she wants? And, moreover, what types of different projects are there, in the first place? Who exactly are the people involved, whom he is going to have to deal with in his or her capacity as client? What are the various options and strategies available for the execution of a given project? What are the respective consequences, advantages and disadvantages?

1.1 Responsibilities of the Client and Project Management

Optimum buildings result from a combination of clear target definition, creative planning and professional execution. Combined with the required expert understanding, any project can become a success if there is professional and timely project management.

Careful and organized planning is just as important as timely coordination of the various measures to be implemented. This reduces frictional losses and creates quality while also saving costs and time. The areas of responsibility for project leaders on the client's side are as follows:

Definition of target specifications for purpose and extent of the construction undertaking

Setup of project structure and contract documents

Decision making and securing of decisions

Ensuring permit compatibility

Monitoring schedule, cost and quality targets

Ensuring financing and marketing

Fig. 1–1 Client's tasks

If we were to unite the execution of all these client's tasks, we arrive at the term known as "project management". A "project manager" or "project leader", on behalf of the actual client or investor, handles this so-called project management.

Among the tasks of the project leader, in the capacity as direct representative of the client, are also expert design of the construction undertaking and representation of executive entity in management and supervisory panels. He/she must be able to make any required decisions and to report on the project's status. In the best-case scenario, the project leader not only presents expert competency but also comes with experience from similar construction projects. Since time restriction prevents a single individual from handling all these tasks while at the same time also

coordinating all the parties involved, he or she requires – for realization of his/her ideas and execution of his/her tasks – a competent team made up of own or outside staff. A "project management team" surrounding the project leader can handle some of the client's tasks like, for instance, monitoring of schedule and cost targets under supervision by the project leader, and can also support him/her in the remaining tasks.

The aim is to take some of the load off the project board, which can then use this time, freed up, to handle those client tasks that cannot be delegated. Tasks that can be delegated, hence, can be handled by project management without taking up the limited time of the project leader. This also lays the foundation for a successful project execution approach. The client may also hand over the entire project management process to an external and trusted contractor of his choice.

1.2 Project Types in Building Construction

While the majority of project types are essentially those of differentiated construction buildings, civil engineering plays a decisive role for transport installations and infrastructure projects, as do mass concrete constructions. In this book, the focus is on building construction projects, whereby, however, the procedures outlined can basically be applied for all project types as long as the necessary modifications are undertaken.

The usual construction projects can be classified according to Figure 1–2.

We need to consider that different project types also place entirely different requirements, of a content nature, on project manager, planners and executing firms. In the case of office buildings, for instance, totally different focal areas and processes apply than, let's say, for a hospital or a sports stadium. For this reason, project management also requires expert knowledge for the respective project types, since otherwise project leadership is doomed to failure.

Private-Sector Office Buildings, Administration

Health Care, Hospitals, Nursing

Training, Education, Research

Public-Sector Office Buildings

Trade Fairs, Congress Centers, Exhibition Venues

Industrial and Production Buildings

Insurance and Bank Buildings

Theaters, Concert Halls, Museums

High-Rise Buildings

Hotels, Resorts, Vacation Facilities

Sport and Leisure Facilities, Theme Parks

Transportation, Infrastructure, Airports

Residential

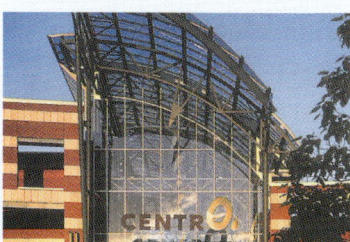

Retail, Malls, Car Showrooms

Supply and Disposal

Fig. 1–2 Project types in building construction

1.3   Project Members

Project members, in essence, can be grouped into four different sections:

– Contractor = client of various types
– Planners and consultants
– Executing construction firms of various constellations
– Licensing bodies and public authorities

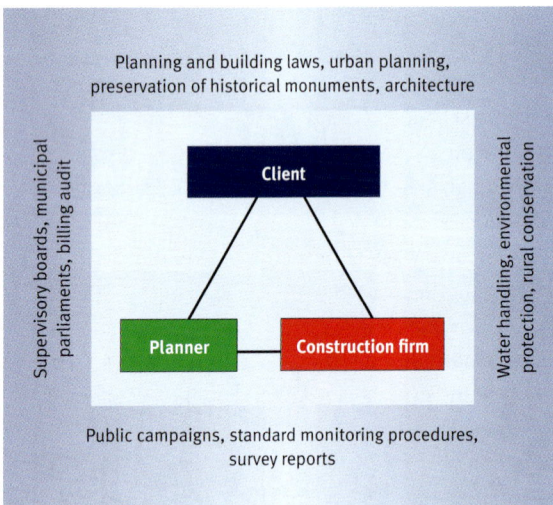

Fig. 1–3 Project members for construction undertakings

The **contractor** or client may consist of an individual or also different types of organization with various target definitions and competencies. He/she/they define(s) the construction task and outline the program, a process for which consultants become involved for larger and more complex projects.

Next, the **planners** convert the program into a location-oriented planning idea and come up with the required calculations, plans and Bills Of Quantities (BOQs). Further, they generally monitor proper execution of the construction services. Their remuneration is according to the German HOAI (Fee Structure of Architects and Engineers – FSAE).

The **construction firms** realize the construction under-taking outlined in the plans and bills of quantities (BOQ) according to VOB or VOL. For this, they either receive per unit remuneration or a flat fee, which was determined from competition with others.

Execution of a construction project may only commence once the relevant **supervisory authorities** have approved the plans and calculations. These approval processes may turn into rather time and cost consuming procedures for larger construction undertakings and this is especially so when there are no valid zoning maps or when these are to be amended. In such cases, there is an enormous increase in the entities involved. Public agencies need to be consulted, e.g. energy supply enterprises, environ-mental protection agencies, and there is frequently also involvement of public campaigns.

We are now going to look at a brief outline of essential variations of clients, planners and construction enter-prises and their different task definitions are schemati-cally reviewed.

1.3.1 Members on the Side of the Client

Client – owner – investor: He/she provides the capital (and, more rarely, a property) for the real estate investment, either from his own or outside sources (financing). For this, he/she receives all returns that remain after deduction of remuneration and fees for construction, marketing and those involved in the running of the project. Here are some typical examples for clients:

– Federal and state administration bodies
– Municipal and county administration
– University and further education administration
– Savings banks, general banks and insurance enterprises
– Social insurance companies
– Industry and commerce
– Construction and residential industry
– Private investors of any type

Project developer – initiator – developer: He/she is the one who has the idea for the projects, goes looking for a suitable property and then prepares implementation including building permit.

Prior to realization he/she finds, if possible, the tenant/tenants and then looks for an end investor to take over the project. As a rule, the project is only implemented once all of these factors are in place. Realizations on spec are associated with high risks and are the exception rather than the rule. The most challenging and decisive service of the lot is that of obtaining a suitable property, and this often happens even before the idea is born. In today's restricted property market, each suitable property, especially when it is also in a top location, is subject to heavy competition. Marketing and envisioned usage is, therefore, often a secondary rather than primary concern. In the undertaking of property search and tenant finding, a project developer generally works hand in hand with a real estate agent or agents. When preparing for planning, obtaining the required permits and supervising the realization of the project, he/she uses his/her own or outside project managers and planners.

The project developer unites supply (of properties) and demand (for real estate) through the development of a given construction project.

Project manager – general manager: The focus in the tasks of the project managers is on optimization of program and planning specifications of the contractor that result from the marketing strategy, in their functional, economical and building engineering aspects, and implementation of the same through planners and construction firms. Jointly with the client, he is to define building standard (what can we offer to the occupants/users?) and then use this as the base for specifying cost and time schedule. Controlling and adherence of this cost and time schedule is something he must ascertain through suitable procedures that are independent from allocation strategy, then the clients is to be kept informed of any deviations and the required decisions are to be brought about.

User – tenant – operator: For usage of the property or building, the user must pay remuneration, so so-called rent or lease. If the case is that of an active business operation, we use the term operator, for instance for hotels, retirement homes, musical theaters or department stores. Users, tenants or operators often hold the function as "quasi clients" especially for interior construction since this is done according to their individual requirements.

A project manager is the representative of the client for all other parties involved in the planning and construction process and only involves the client for important decision making, which he or she has already prepared accordingly. His/or her respective authority is contractually specified dependent on the management capacity of the client. They may range from mere management support all the way through far-reaching representation entitlements.

1.3.2 Possible Types of Organization for Planners and Consultants

The number of planners and consultants involved in a given project has continually risen over the years. This is the result of several different developments:

– Complexity of technology for and in buildings has increasingly risen, which is the reason for ever more specialized expert teams having been formed.
– As a direct result, permit procedures and approvals have also become increasingly more complex, meaning that additional expert assessors are now employed on all sides.
– When signing their contracts, many clients increasingly attempt to pass on risks to the contractors. Also, investment projects require clear assessment of returns, unlike the previously much more common auto-use projects, and this has resulted in construction costs coming under more and more pressure. Entrepreneurs have reacted to this with claim management. This situation has led to increasing use of attorneys on both sides.

Individual planners: These – freelancers as a rule – architects and specialist planners for structural design, technical insatallations, building physics etc. take a step-by-step approach to implement program and standard specifications of the client/project management according to the service outlines of the Fee Structure for Architects and Engineers (FSAE) and transform them into planning concepts, permit and execution plans and performance descriptions, according to which the executing firms then work. To this end, they generally undertake local construction supervision (Object supervision, building- and object supervision, building services equipment).

In many cases, executive planning is nowadays undertaken by the construction firms but these, frequently, employ outside specialist engineering bureaus for the task. Aside from formwork and reinforcement plans for reinforced concrete works, this also relates to execution planning of the concrete buildings as well as technical trades and extensions. In the case of pre-fabricated buildings, the entire planning services are generally undertaken by the executing construction firm.

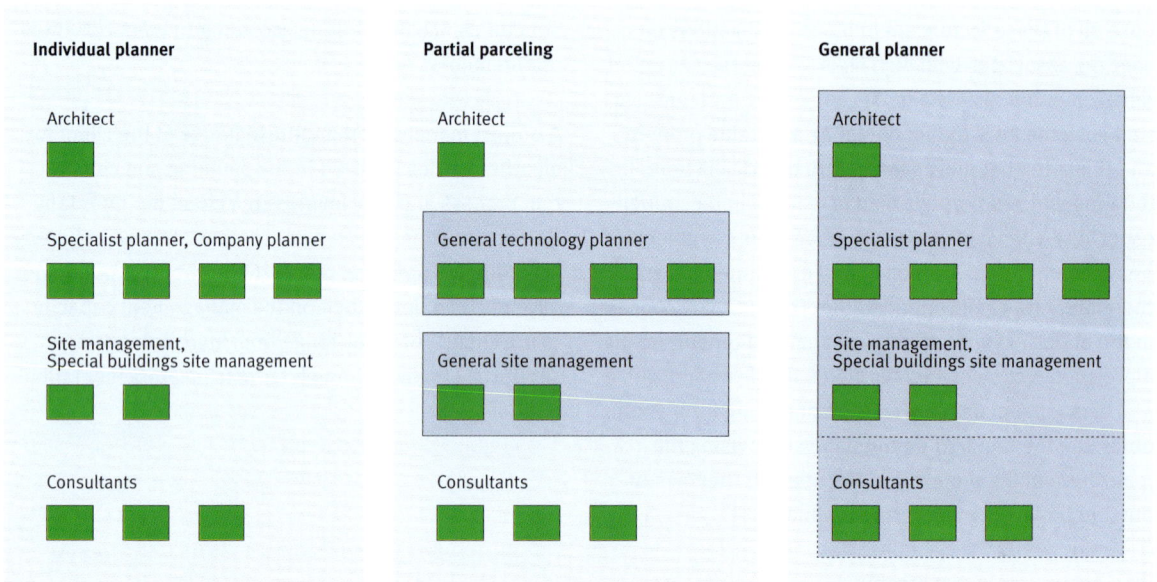

Fig. 1–4 Possible constellations for planners and consultants

On account of specialization, there are nowadays, as a rule, separate consultants for kitchen planning, conveyor and transport systems, for protection and smoke extraction, low voltage systems, safety engineering etc., who also need to be integrated into overall planning.

Content coordination for the individual persons or entities involved in planning is with the architects who, however, are often overloaded, especially when it comes to major projects. As a result, significant coordination services, also of a content kind, fall back onto project management in the end.

Partial parceling: Massive coordination requirements for the individual expert trades as well as the need for integrative planning for sustainable buildings (especially heating, ventilation, sanitary, electric equipment, building physics, facade construction, roof) have resulted in larger suppliers offering these services as part of general expert planning with in-house coordination. This reduces interface problems.

The same applies to parcel acceptance of object supervision for buildings and technical engineering as part of the general site management.

General planner: An increasing number of persons and entities involved in the planning process on account of continuing specialization in the individual specialist sector has led to the use of general planners especially in the areas of industrial and facility construction. Their scope encompasses all planning services, including architecture, either with their own staff or sub-planners working on their behalf. In this, they are the solely responsible contact person for the client and frequently also handle project control on the side of the contractor. General planners are often limited planning companies from architectural firms, since they are often on their part assigning parts of the tasks to freelance individual planners but then coordinate these in-house.

One problem of general planning, frequently, is that highly qualified specialist firms do not generally like to become part of this organization as sub-contractors only and this is something that applies, especially, to

renowned architects. Further, management competency is often not sufficiently advanced to a level as would be required to place adequate focus on process optimization and economic aspects.

1.3.3 Possible Types of Organization for executing Firms

Organizational structure of firms executing the construction operation can be designed in very different ways, depending on the target definitions of the client. If he or she wants to have the largest possible influence on design and undertake step-by-step design decisions, he or she needs to remain as contractually flexible as possible and only ever assign as many tasks as are required for the current phase.

If, instead, he or she desires flat fee and guaranteed deadlines primarily, one single contract partner would be the strategy of choice. This is an overview of the various options that exist:

Sole entrepreneur (Shell construction trades, fitout trades, engineering trades, interior design and equipment, outside areas): The sole entrepreneurs handle definition of the various trades for a given project according to the plans and tender documents of architects and specialist planners. For specific trades (especially facade, engineering and fitout trades with factory assembled components) they create their own assembly and trade-shop drawings.

As a rule, they receive remuneration according to individual items and mass calculations, applying escalator clauses. Alternatively, fixed fee agreements are to be recommended since they often only require a small additional cost to transfer inflation risk to the side of the entrepreneur.

Partial general contractors: For very difficult and demanding construction undertakings, recent times have frequently seen a uniting of the trades, with many interfaces and mutual coordination demand, for so-called partial GC or component GC (parcel assignment).

Typical examples are the engineering trades "Heating – ventilation – sanitary – sprinkler – gas supply" or "suspended ceilings – dividing walls – wall units" in office building construction. These trades then form a work unit, which as partial GC for these trades gets the overall assignment. Processing is similar to the variety with individual planners and individual trades since here, too, everyone involved needs to receive their assignments at once. Therefore, it is possible to work at the same efficiency level and with only marginally restricted flexibility.

Another example are the trades "roof – facade". For this combination, a rain-proof-building can very quickly be worked out with the respective firms and then be assigned to the same, without the rest of the planning presenting the degree of detail that would be required for GC assignment.

This procedure, however, does carry some risks. For instance, these types of assignments require in-depth knowledge on the side of project management about the market and planning contexts.

General contractor: The development of a general contractor idea has resulted from the desire of the client to have "guaranteed" costs and times schedules while at the same time reducing his or her own involvement. As a rule, the GC is a building contractor himself or herself – frequently for shell construction works – while he or she outsources the other services to subcontractors. A general contractor is responsible for the ready-to-move-in construction of the building, based on planning documents and service description. He or she absorbs the price risk that comes with outsourcing building services to subcontractors and signs a GC contract with the client, at fixed price and for a specified schedule

In this, however, there are frequently certain risks that are either rejected or only taken on at excess prices.

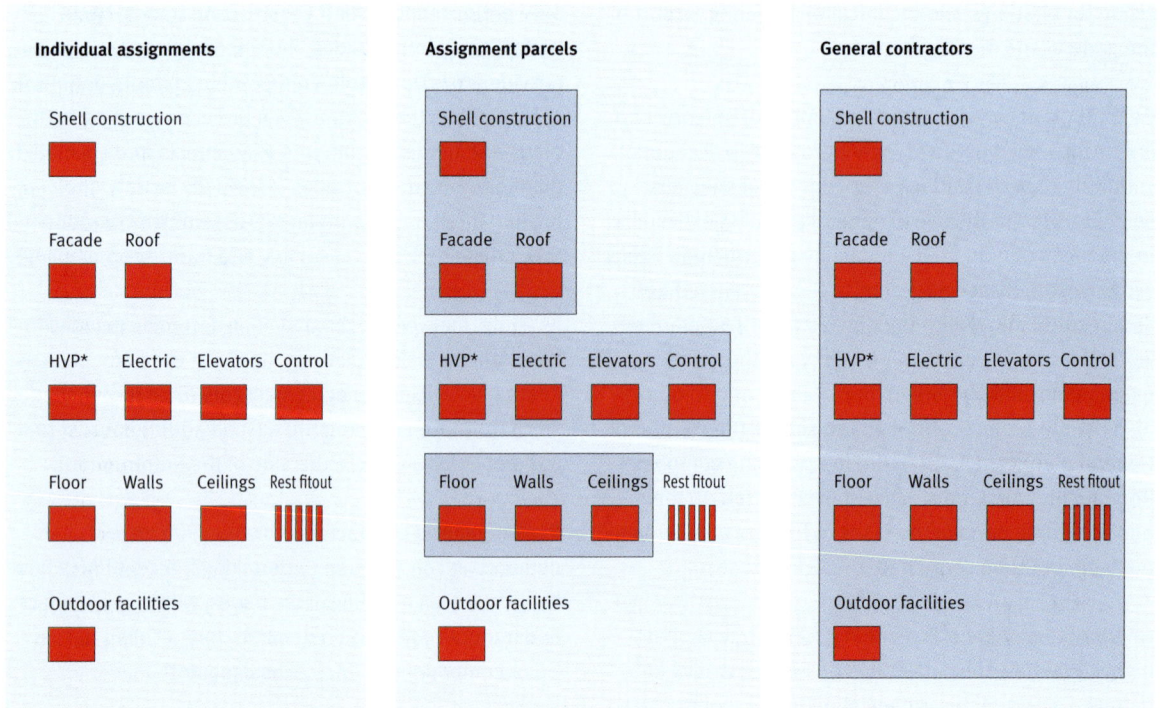

Fig. 1–5 Different assignment strategies during construction execution

★ Heating, Ventilation and Plumbing

Examples for this are as follows:

– Building site risk: Support capacity, containment of water, soiled subsoil, polluted ground water
– Risks resulting from building permit process; especially fire protection stipulations, stipulations of trade supervisory authorities, which are frequently up to the discretion of these authorities and cannot be determined in advance.

A general contractor does not need to absorb the risk for construction permits, interim financing, for lease, rental or sale of the object. There are several options for interfaces concerning planning and construction preparation:

– Complete provisional planning with individual bills of quantities (HOAI phase 1 to 7): In that case, planning is completed all the way to execution planning for all trades and regular bills of quantities are also created for all the trades. In this case, a GC functions only as a purely executive firm without exerting influence on the planning process itself.

– Reduced provisional planning with room manual and construction work log book: As tender documents, permission and execution planning by the architect is available but this needs to be taken more as system planning and depiction of essential architectural details. Further, there is a complete room manual as well as a detailed construction and quality description.

For structural design and engineering equipment, preliminary planning and parts of the design are sufficient, supplemented by the corresponding construction and quality descriptions. Other planning services are supplied by the GC and approved by architects and expert engineers. This is the usual procedure for individually planned and architecturally sophisticated buildings.

– **Design and construction description (building specifications) for simple construction tasks:** For more simple buildings, qualified preliminary design and construction description according to trades is sufficient. In that case, a GC has the most optimization possibilities, which is reflected in a cost-effective bid

but also in constant fight for the quality desired. In that case, the GC is already asked to become involved in the course of permission planning that the architect is to undertake for the contractor. The GC takes over architect's planning from the point of execution planning, whereas the PR usually will grant the architect artistic supervision.

– **Anglo-Saxon method:** The contractor supplies complete planning and building specification together with the tender documents. These plans and "specifications" contain all planning information in detail but they do not constitute well worked-through and coordinated execution planning, as is the case for the HOAI. Rather, it only shows what the contractor later "sees". The precise manner of implementation is up to the general contractor. He or she, hence, is also to create execution plans for the architect, structural design and building services equipment.

This frequently constitutes a source of misunderstanding between German clients and British or U.S. architects since the latter often tend of assume that the entrepreneur will handle this type of coordination. Therefore, they tend to concentrate on visual appearance of the building and on all the details that appear important to them. How these are to be realized is something they often place less emphasis on – this is a problem that needs to be solved, period. They are, after all, the "designers" and, as a rule, are being used in that capacity. When a client does not work with the GC principle, as is frequently the case, for instance, in Scandinavia, a large engineering firm takes over execution planning in the sense of the HOAI.

Coordinating general contractor – property developer:
The coordinating of general contractor or property developer already enters the project at the start of planning. He/she takes over planning and construction of the property, financing and then also handles leasing or renting out or sale: he/she bears the economical risk until the point of operation-ready transfer to an investor, including interim and final financing. This model is typically applied for residential apartment buildings or for office buildings that are later to be rented out.

1.4   Alternative Implementation Strategies

Many clients in their time have had to realize that having a number of parties to the project can, in absence of coordination by a strong and experienced partner, lead to many disadvantages and annoyances.

For this reason, there is an increasing tendency towards generalization in different areas with the aim to use such generalization to obtain better control and monitoring of the construction project. This applies for a so-called "general management" on the client's side as well as for the "general planner", the "general contractor" or the "coordinating general contractor". Aside from specific, special cases, all building construction projects run according to a certain scheme, more or less, where by project duration can widely fluctuate according to client, organization and dependency on permission procedures.

As a rule, the largest deviations in such a process occur as a result of different constellations in conjunction with the triangle **"client – planner – property developer"**, which the project manager needs to be familiar with.

1.4.1   Building with Sole Entrepreneurs

The traditional way of handling a construction project is organization via individual planners and contractors. The biggest advantage here, without doubt, is that you can select and obligate for planning as well as for the individual trades the respectively best company, according to price, quality and output. Another advan-

tage is a high level of flexibility. Changes in planning can be adsorbed much better than is the case for a blanket contract because services are only called on in accordance with planning and construction progress.

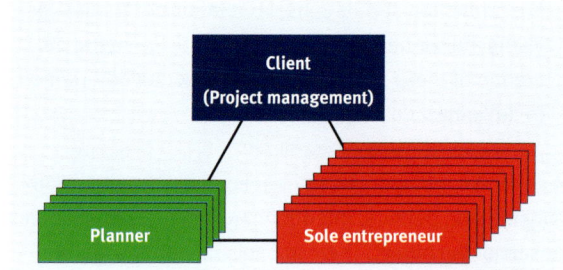

Fig. 1–6 Organization with individual planners and sole entrepreneurs

The entire coordination, however, hereby is with the client. If he decides to involve project management for the coordination, maximum service range is called for, which means a lot of effort for the project manager. At the same time, however, this is the ideal constellation for professional project management since the most extensive influencing options are granted in this way. For the client, this results in optimized economizing, functionality, quality and design.

The process itself is characterized by heavy overlapping of planning and construction execution. This overlapping on one hand opens doors for employing state of the art technology and construction procedures while on the other it also comes with the risk of costly planning amendments. If professional project management knows how

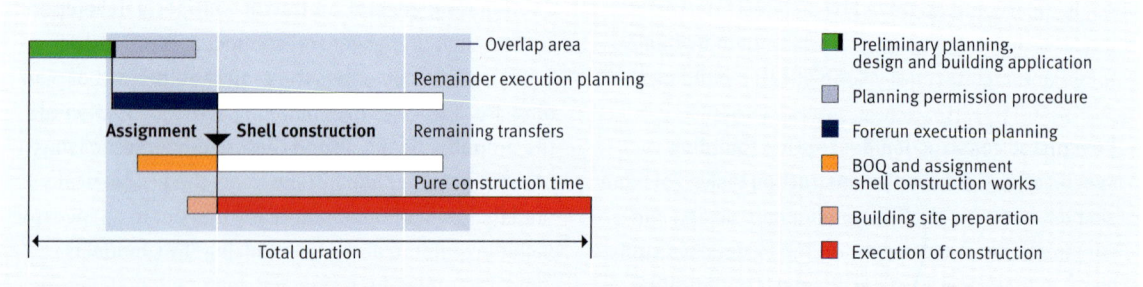

Fig. 1–7 Conventional process with overlap planning and building construction

to handle this, this process, especially for difficult projects, allows for shortest project duration at highest quality thanks to its great overlap opportunities for planning and execution.

1.4.2 Building with Parcel Allocations (Part-GC)

When building with parcel allocations, there are numerous advantages to project management in the case of larger and more complex construction undertakings. For instance, the knowledge of specialist firms can be incorporated early on into the project execution process, which – as a rule – leads to significant cost savings with simultaneous quality improvement.

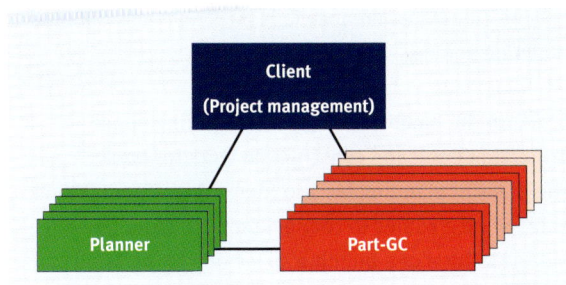

Fig. 1–8 Organization with individual planners and parcel allocations

In that case, project management does not need to look after coordination tasks within the Part-GC trades and thus is able to spend more time on optimization of functioning, economic aspects and quality of the construction undertaking. This is also an opportunity for frequently improving of such aspects as schedule processes and logistics at the building site, thanks to further reaching pre-assembling.

1.4.3 Building with General Contractor (GC) = Turnkey Construction

Project execution in this constellation spares the client/ project management coordination tasks at the site but requires a large amount of coordination effort during the preparation stage. The guarantees of the GC can only

be relied on whenever there is mature and coordinated planning at least until the point of building application.

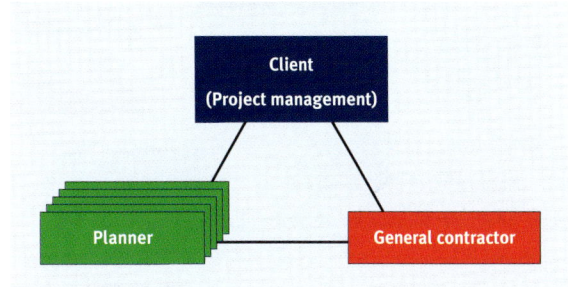

Fig. 1–9 Organization with individual planners and general contractor

In this version, the GC frequently takes on individual planners also for execution planning. Project management hereby fulfills the function of controller and trustee of the client. He/she needs to monitor quality and payment plans, track addendums and inspect them and, to this end, also keep in mind the agreed-upon schedule with inspection dates. On one hand, this means that effort is less because there is no need for detailed schedule coordination with subcontractors, or for monitoring individual invoicing. On the other hand, however, the entire monitoring process becomes more controversial since there are often more addendums due to changes in planning or more in-depth planning, something that any GC will take as a good reason for significantly upping his/her fixed price flat fee – whether this is justified or not. To defend against addendums that are not justified, the project manager needs to delve deeply into the matter, both concerning costs and schedule sequence.

In net terms, effort during preparation and building construction may be less by 15 to 20 % (expecting full performance capacity), while the client receives cost and schedule guarantee as well as the security that these will also be implemented thanks to the efforts of project management. This security, however, is going to cost more than handling the project via individual assignments.

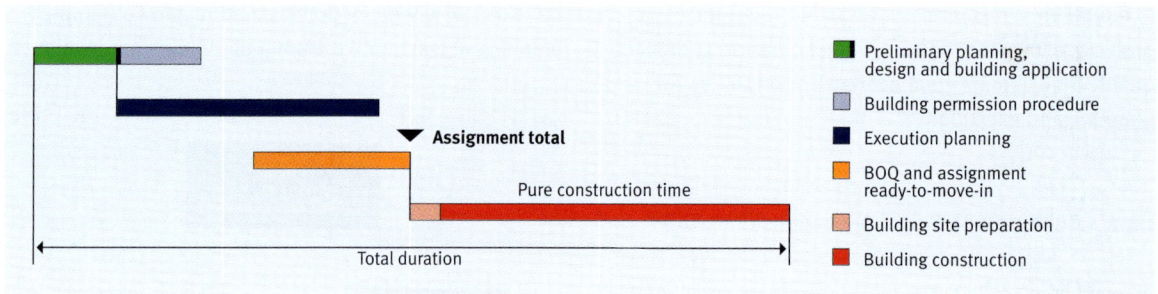

Fig. 1–10 Project duration for GC on the basis of execution plans

When building with general contractors, there are also several different varieties as far as the time of inclusion and present planning status is concerned. Contractually, we distinguish between three varieties:

a) GC on the basis of execution planning:
For this variety, we assume that the executive plans are essentially present, at a scale of M1:50, and that all services to be performed have been precisely defined in quality and quantity terms (American method). Advantages are clear calculation documents and good comparing options, together with a high degree of cost security as long as there are no subsequent changes.

There are, however, little to no rationalization options for the general contractor, which is disadvantageous, as is the long project duration compared to convention execution. Due to lack of overlap between planning and execution, duration is about 12 months longer. Since the GC still needs to calculate an add-on for cost and schedule guarantees, one might as well, or even with more

advantage, take the option of individual assignments.

Overall, this method is not very suitable for execution with GC since it leaves him/her little leeway, which can only be created when once falls back on a less detailed planning stance.

b) GC on the basis of design with building space utilization book and construction work logbook:
The basis of planning M1:100 (design) is generally suitable, when it comes in connection with a fundamental outline in form of a building space utilization book and construction work logbook, to increase rationalization leeway for the GC as well as adequately safeguarding the client's interests.

In this, there is decisive emphasis on schematic design quality, which needs to be thoroughly worked through as construction planning and precisely outlined in the building space utilization book and construction work logbook (for this, see subsequent details that are to follow).

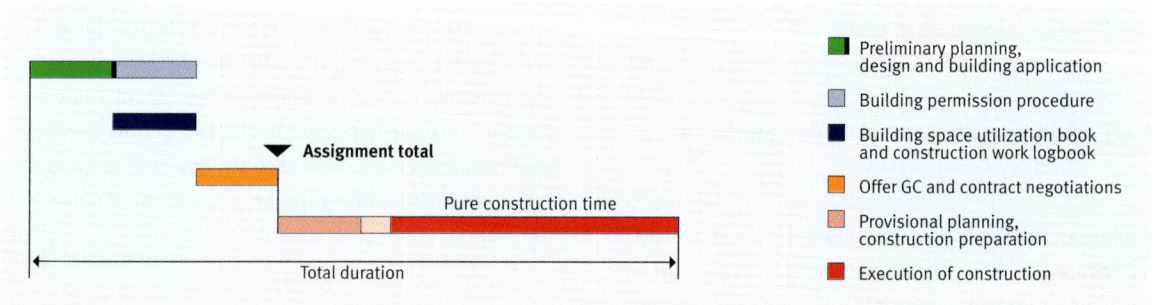

Fig. 1–11 Project duration for GC on the basis of design with building space utilization book and construction work logbook

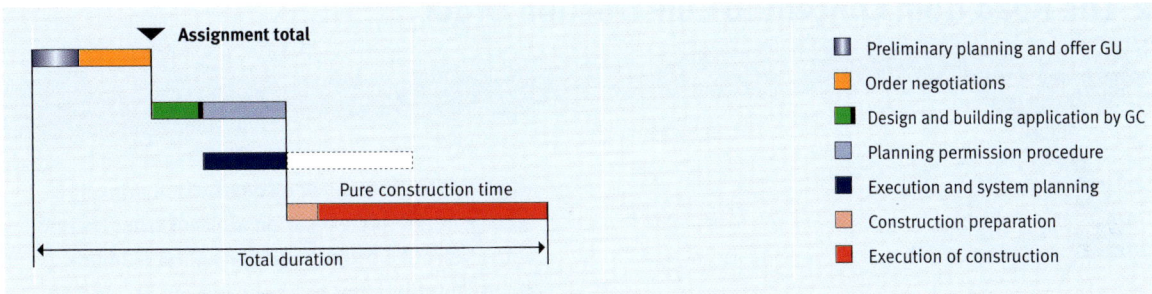

Fig. 1–12 Project duration for GC on the basis of space and function program

Project duration for this procedure, owing to early assignment to the GC and his/her options for rationalization of the process, is about the same as for conventional processes with the advantage of cost and schedule guarantee. However, subsequent changes have a more grave impact for complex projects than they do for the GC1 variety, since the basis of cost definition is only outlined in principle but not in detail. For regular projects, however, this is a suitable process.

c) GC on the basis of space and function program:
If the focus of an investment project is on as cost-effective erection of a building as possible (without too many demands on long-term quality) while keeping investment costs low, it is advantageous to move up the assignment time even more.

In that case, the GC can already take own construction undertakings into account during preliminary planning since he/she is not furnished with plans but merely with a space and function program as a foundation for his or her bid. Most frequently, the lease or rent to be obtained serves as a base to calculate maximum possible investment, which is then negotiated ad lib. This process is also suitable for a type of combined planning and realization competition that, however, places high demands on the client's project management in terms of assessment.

Project duration for this process is comparatively short. However, the client has basically no options to exert influence following the assignment, in case that he/she does not wish to lose advantages gained.

1.4.4 Building Partner Model

In the building partner model, firms of the various trades – usually project-related – unite into a joint executive firm where each is a partner to approximately the same extent as is represented by the proportion of his/her trades. The offer is worked out jointly and put in via the person in charge, partially also with schedule and cost guarantee. Success, here, is essentially dependent on the skills and experience of the person in charge who also needs to handle the entire coordination.

1.4.5 Coordinating General Contractor

A coordinating general contractor reduces the client's tasks to pure controlling of the contract content agreed upon and the cost plan. Coordination is only undertaken directly anymore between client and coordinating general contractor. Since the latter also has the architects as contractors on his/her side, no significant influencing or optimization is possible anymore once the fundamentals have been specified. Essentially, this version is used for pure investment projects that are being marketed via closed real estate funds. Aside from payment of the sales installments, the clients are not directly involved in project procedure, which means that their monitoring regarding conformity with construction progress and constant quality control becomes all the more important. As a rule, there is then a downstream general contractor.

2 The Road from Concept to Construction Order

The basis for success of a construction project, as far as content and execution are concerned, is laid at the start of a project and during the planning stage. Anything that has been left out or handled wrongly during this stage can only be repaired to a limited extent during building construction. Unfortunately, time is very frequently wasted during that stage, which then leads to hectic scenes during the execution of the project and to chances that were plain and simply missed. For this reason, we are going to look at this particular stage in detail.

2.1 From Construction Task to Planning Idea

The stage between clarifying the construction tasks and actual planning idea is of enormous significance for quality level of a project. Decisions made have far-reaching consequences for architecture and urban planning as well as economical and ecological quality of the buildings. Roughly put, the procedure can be described as follows:

For the client, a given project begins with the definition stage, where the construction tasks are outlined and the project goals specified. Generally, at this stage, there are no services yet according to FSAE.

Subsequently, there is the search for a suitable plot and analysis of the corresponding fringe conditions.

Afterwards, the desired project volume can be determined in a more concrete manner and depicted in a space and function program.

Via a planimetric model, the gross areas to be expected and cubic content are determined and expected costs deduced. If the return on investment ratio is acceptable and cubic capacities in accordance with the plot and the expectations of the community, it is now possible to either involve an architect for the creation of a planning idea or to launch an architect competition.

2.1.1 Construction Task Clarification and Definition of Project Targets

At the start of this stage, there is a question: "What, exactly, are we trying to achieve?" The more you look at this question in-depth, the more difficult it becomes to answer it. All this despite the fact that the question "What do we wish to avoid?" is just as important and, moreover, easier to answer. These are some examples for the usual problems encountered in office and administration building construction:

– Lack of space due to growth
– Problems on the organization side
– Lack of communication
– Problematic work areas
– Insufficient productivity
– Insufficient presentation of the company to the outside

At first, there must be the basic distinction of whether one wishes to build for oneself and/or one's own company or whether the project at hand is a pure investment project that is to be marketed.

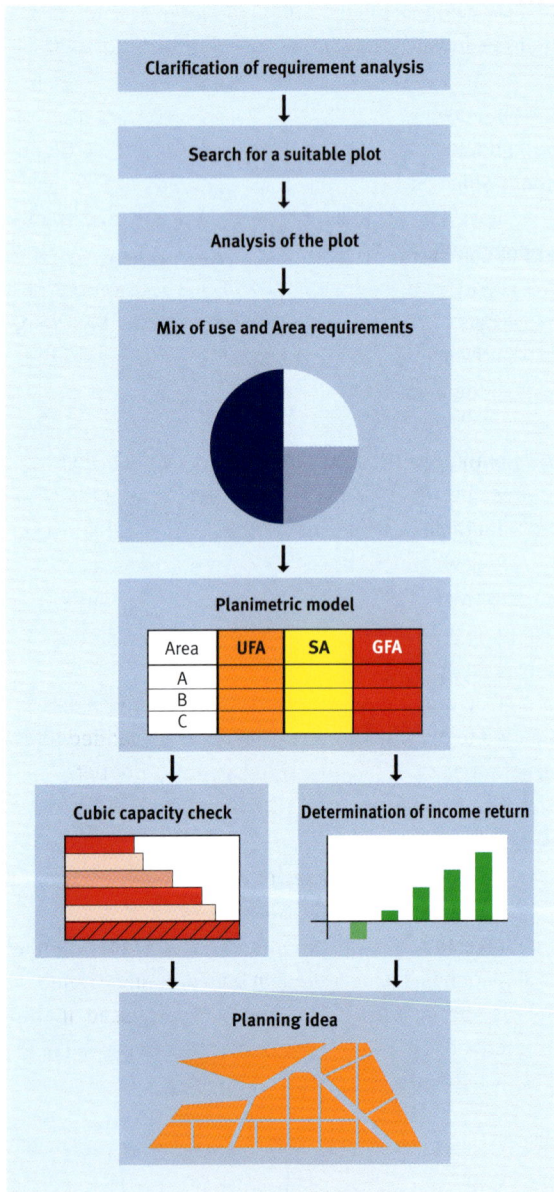

Fig. 2–1 From construction task to planning idea

Depending on these premises, the following target definitions are to be weighed:

– Return on investment and productive investment
– Improvement of productivity and use of synergy effects
– Functionality and staff satisfaction
– Improvement of corporate identity and corporate culture

2.1.2 Search for a suitable Plot

One of the most difficult and time-consuming tasks, certainly, is the search for suitable plot in a location as good as possible.

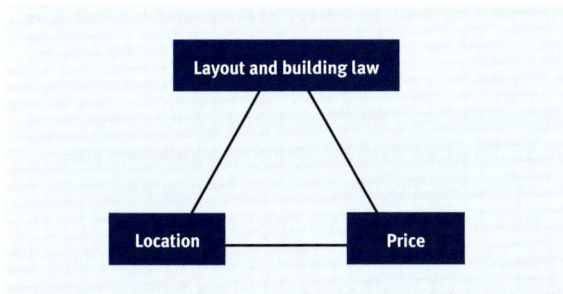

Fig. 2–2 Plot parameters

In this, the price of the plot plays a decisive role when calculating return on investment. It depends primarily on location (Center or fringe location, traffic connection, shopping possibilities, public institutions), plot layout and available construction rights (Development options).

2.1.3 Plot Analysis

With the increasing amount our congested urban areas are built up, anyone looking to build is going to encounter great difficulties finding a suitable commercial plot, whether already built on or vacant. Sufficient size, excellent location and a developed terrain as well as a functional infrastructure are all required. If one has

finally succeeded, such a plot can still come with a number of unpleasant surprises. Especially in the case of congested urban areas, many dangers lurk, so that professional location analysis is important as part of the plot search process.

Even when a plot is already available, such location analyses, in slightly adapted form, are to be urgently recommended to be on the safe side of planning. We are now going to touch on some essential criteria that are to be looked into during such a location analysis.

Building law: At the very least, a land development plan should be available in the case of congested urban areas, which allows for building on the plot concerned. This, however, still does not provide any type of certainty for development potential, which can be provided only by a valid zoning map. Such a zoning map, however, does not always come only with advantages, especially when, for whatever reasons, there were very restrictive guidelines at the time of its creation.

Town planning: While the urban planning authority's desires may not necessarily be binding for later planning, it is nonetheless important to know for the later process just along what lines those in charge of this area do actually think. This, primarily, is about incorporating a new project into its environment.

Infrastructure: For large projects, it is urgently recommended to have a traffic report done by a qualified firm prior to the actual beginning of planning. Adequate access to energy and water for the building area, as well as to availability of disposal options, are a further important component of the infrastructure.

Mortgage on land: Only when the development potential of the plot has been clarified, in both size and extent, can the specified plot price be properly assessed. In the end, it needs to be looked at per m2 of productive land/ effective surface and/or work place.

Development costs are another significant cost factor when looking into the purchase of a piece of land. In this, we generally distinguish between connection costs that relate to consumption – such as for electricity, heating media like gas, and water – and general take-over costs as are usually agreed upon nowadays for such things as access to traffic infrastructure and waste water disposal. If a plot in the urban area is already built on, e.g. if there are still some buildings present, then demolition works can very well become substantial, especially if there was an industrial operation there before that has produced products with contaminated remnants.

Further, re-positioning or replacing of existing lines can become a significant cost and time factor since, as a rule, new routes need to be found that may require their own permission processes. Especially when it comes to lines by Deutsche Telekom AG or other network providers, or the Deutschen Bahn AG (German Rail), lengthy negotiations are the rule rather than the exception. Often as a way out, the choice is made to keep the lines as they are and hence deal with the problem of the resulting need for integration into the new building that is to be planned. This is frequently considered the smaller evil, although significant costs are also associated with this.

There is a need for especially thorough research at local authority real estate offices into whether there are any additional existing ownership, use or access rights for the property. Such burdens, as a rule, turn out to be highly dangerous when one wishes to deviate from existing planning law and a development plan amendment procedure is envisioned.

Building site: An essential influence, especially on basements and the foundation of planned buildings, is the position of the ground water, traffic routes and piping routes.

Old burdens: If, during the research, there is suspicion of contaminated soil, a chemical analysis is urgently recommended. If it turns out that there is a need for rehabilitation, the perpetrator concept generally applies. However, principally, the owner is responsible for the rehabilitation undertaking.

While transport problems in congested inner cities and also recycling possibilities are a primary consideration for regular excavation material, the primary considerations for contaminated soils are the different rehabilitation options. Currently, we are to assume that the transport of highly contaminated soil can run up to 1.000,– €/t, while a possible decontamination on location results in costs of 20,– to 600,– €/t.

The sooner one deals with the problem, the more it is possible to limit the impact on later planning and realization. Hence, as a rule, it makes more sense to partially forgo development of the site and to leave the soil where it is. To do this, the respective current laws need to be checked concerning disposal obligation for the entire site.

Ground water level, catchment area: Nowadays, ground water level is a decisive factor for the development potential of a plot in the basement areas, for instance underground garage, control room and storage area. Building or laying foundations, when it is done in ground water, causes significant additional costs, meaning that prior to buying the plot one needs to attempt to find out ground water level in the plot area from the authorities. The final certainty, of course, is only on the basis of an official soil assessment, which however is too expensive and time consuming prior to purchase of the property because certain statements about the ground water can only be made by means of a perennial well. Special care must be taken whenever the plot is in the region of a catchment area.

Environmental stipulations: Environmental protection agencies, in essence, contribute the following factors to the development process:

– Preservation of air exchange corridors
– Reduction of emissions
– Protection of water bodies
– Tree protection
– Avoiding final covers

Over the course of a policy discussion with the relevant authorities, the superseding visions for this area can be clarified.

Preservation of evidence procedure: If the property borders directly on other buildings, commuter train, subway routes as well as supply and waste disposal lines, in if in-depth construction works are envisioned for the area of the property borders, then preservation of evidence procedures, even if they are partially costly, are to be recommended and probably even obligatory. The costs for such preservation of evidence measures, especially in the inner city area where often 100 % of the property is equipped with an underground level, can be substantial. These costs also place an additional load onto the total property costs. Risky neighbors, in this aspect, are especially routes for subway and commuter trains due to the inaccuracy of rail layout, as are large sewer tunnels.

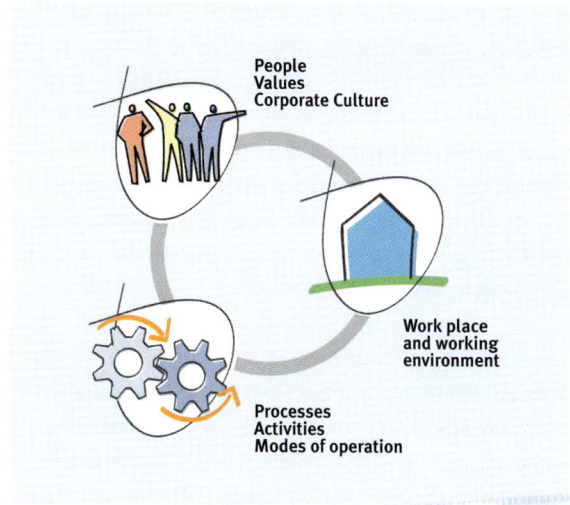

People
Values
Corporate Culture

Work place
and working
environment

Processes
Activities
Modes of operation

Fig. 2–3 Criteria for programming

2.1.4 Programming

Looking at the example of a new office building, it is quite easy to recognize criteria that affect programming:

In order to allow staff to work both in a motivated and productive manner, work organization and building structure should be supporting the mode of operation in the best possible way.

Mode of operation, organization and rooms are subject to constant change. Few work and space concepts correspond to these requirements for a future-oriented and work environment that is supportive of whatever procedures.

Communication, sharing knowledge and cross-department cooperation are becoming increasingly significant

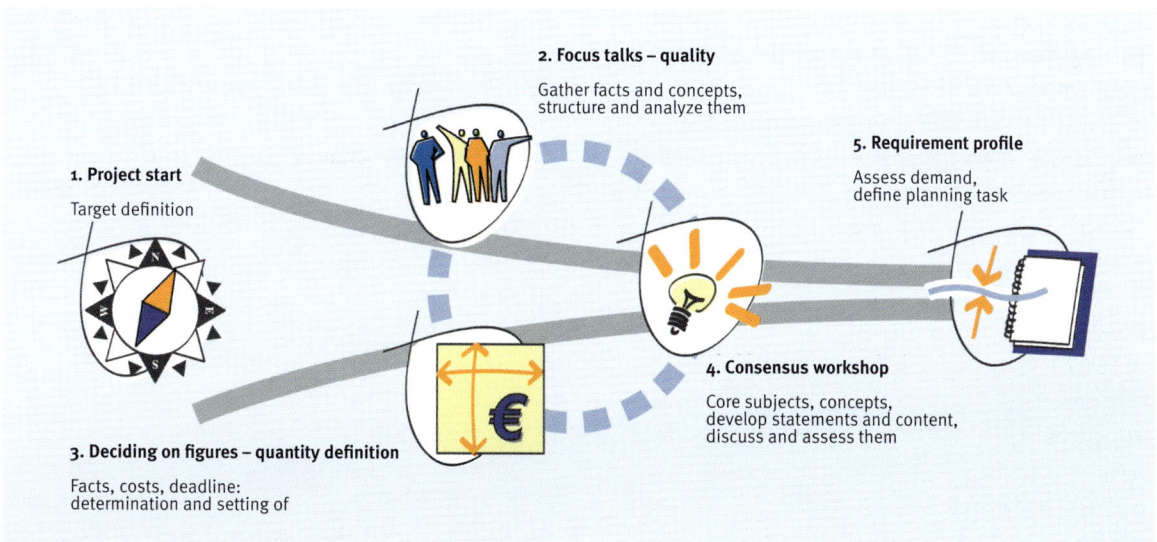

2. Focus talks – quality
Gather facts and concepts, structure and analyze them

1. Project start
Target definition

5. Requirement profile
Assess demand, define planning task

3. Deciding on figures – quantity definition
Facts, costs, deadline: determination and setting of

4. Consensus workshop
Core subjects, concepts, develop statements and content, discuss and assess them

Fig. 2–4 Methodical approach for programming

and can be optimally supported through both building structure and working environment.

Building structures that can hold up to the future ought to be flexible and it should be possible to adjust them with little effort to new requirements. The new conception of a building or of existing spaces is always also a chance to shape the future. High quality and sustainable planning, therefore, is essential.

The required rooms are divided up according to individual useful areas in a surface range, including statements regarding dimension between axes, room dimensions as well as room and floor height. Specifications regarding use are supposed to illustrate the compressed target ideas of the users and to summarize the entire space and function program in a kind of overview that suffices as the basis for initial design ideas.

Using the Stuttgart Trade Fair Center as an example, you can see the specifications for use outlined in the graphs found in Figures 2–5.

The requirements of the new Trade Fair Center, for example, concerned two separate areas:

– Exhibition and Delivery
– Visitor Control

The basis here was a requirement for "short routes". It was also essential to establish a separation between exhibition/delivery traffic and visitor control. The parking zones for these two groups also needed to be separate when it came to the layout.

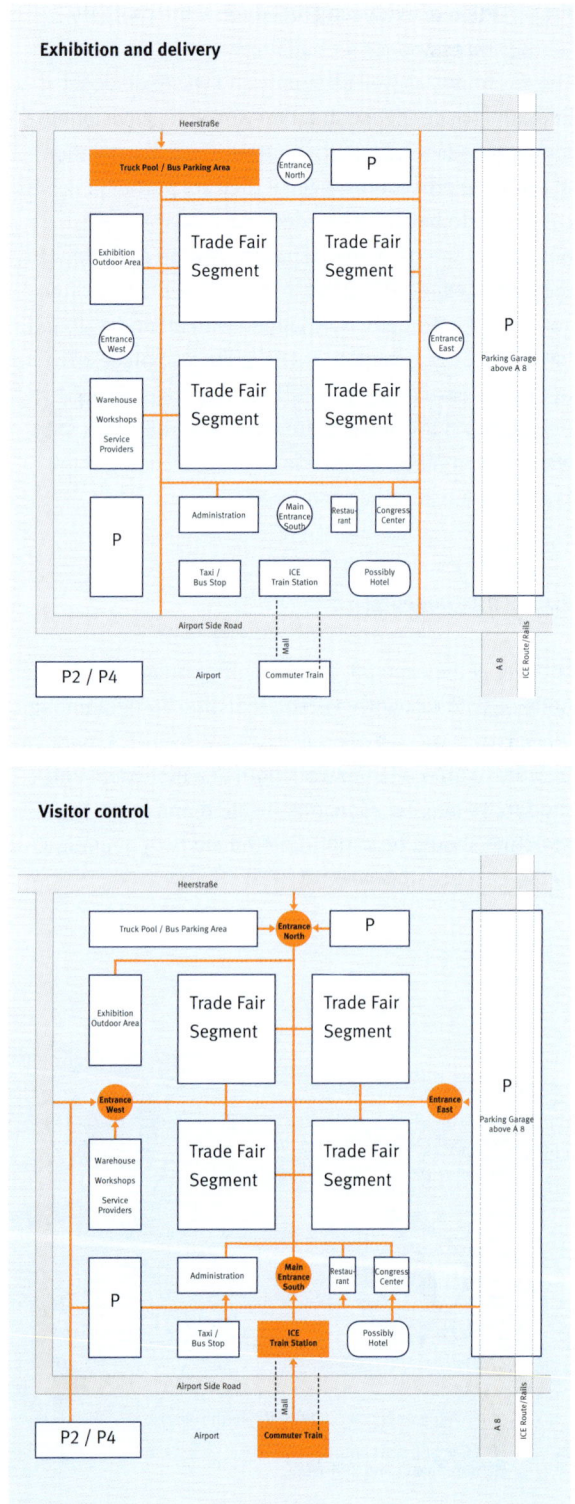

Exhibition and delivery

Visitor control

Fig. 2–5 Functional concept (Stuttgart Trade Fair Center)

Finally, an order of favorites is arrived at and a first prize-winner announced, whose design is then recommended for execution through the investor.

Competitions can be held according to PGC or through assigning several architects.

In the case of competitions that are held according to PGC, which also involve European service directives as a whole, the boundary conditions for architectural conditions have been specified; among others, one presumes an anonymous process. The only exception – a cooperative procedure according to PGC – is suitable especially for complex projects where the solution is determined jointly by the competition holding entity, the award jury and participants by means of an exchange of ideas. Competition panels of architectural chambers, and in the case of public clients also additionally public authorities, are there to monitor that the rules are adhered to.

When several architects are assigned at once, the options are either a cooperative or an anonymous process. In the case of the former, architects and client, so to say, work on the basis of direct coordination. If the process is an anonymous one, on the other hand, a neutral award panel decides on the level to which the submitted works qualify; generally, the client will follow the recommendations of the award jury but he or she is also granted the option of assigning the task to one of the other award winners.

Currently, the version of a combined competition is under assessment in the PGC as a special procedure. This combination of planning and construction service aims at lowering construction costs.

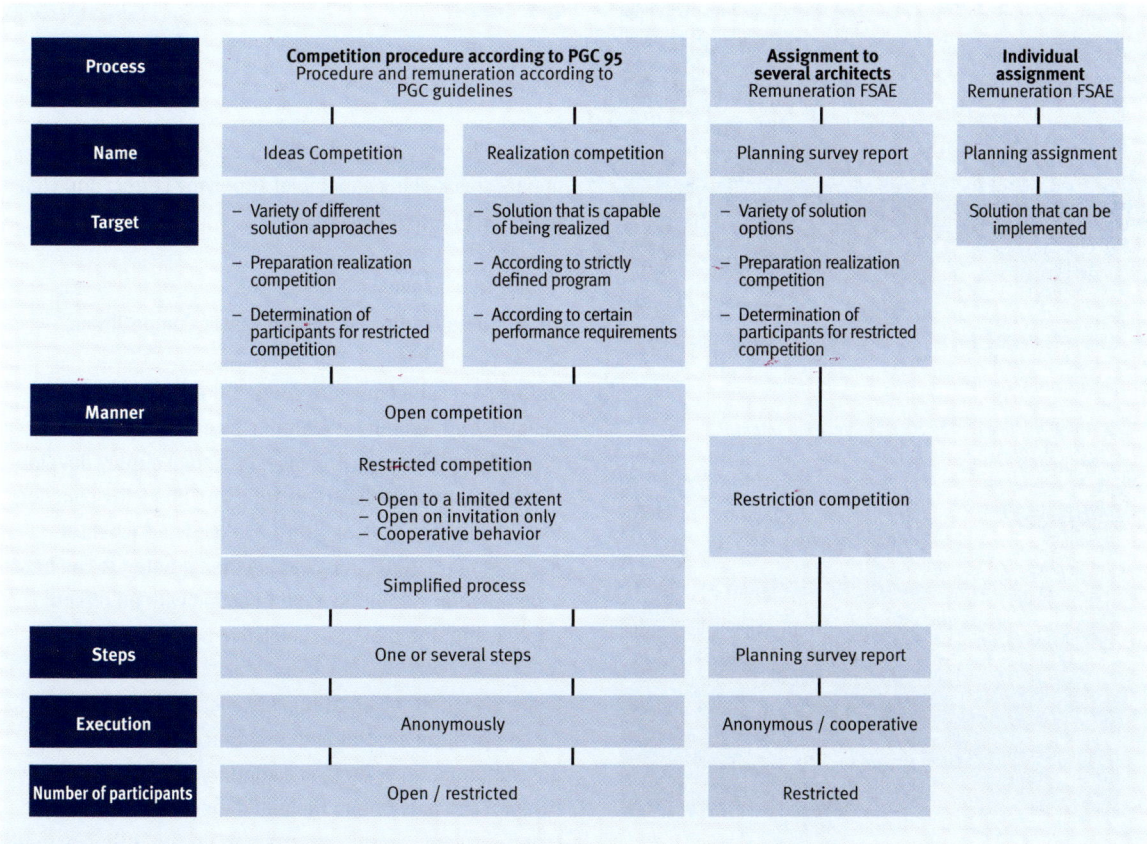

Process	Competition procedure according to PGC 95 Procedure and remuneration according to PGC guidelines		Assignment to several architects Remuneration FSAE	Individual assignment Remuneration FSAE
Name	Ideas Competition	Realization competition	Planning survey report	Planning assignment
Target	– Variety of different solution approaches – Preparation realization competition – Determination of participants for restricted competition	– Solution that is capable of being realized – According to strictly defined program – According to certain performance requirements	– Variety of solution options – Preparation realization competition – Determination of participants for restricted competition	Solution that can be implemented
Manner	Open competition			
	Restricted competition – Open to a limited extent – Open on invitation only – Cooperative behavior		Restriction competition	
	Simplified process			
Steps	One or several steps		Planning survey report	
Execution	Anonymously		Anonymous / cooperative	
Number of participants	Open / restricted		Restricted	

Fig. 2–9 Possible alternatives to involving an architect

In practice, it can often become a difficult task when determining a competition winner to find the correct consensus between material award judges as representatives of the investor (goals: functionality, economical aspects, marketability) and subject judges, represented by the architects (goals: urban planning, architecture, ecology aspects). In essence, a good result depends on one hand on the requirements for the competition being sensible and professionally outlined and on the other on the preli-

Fig. 2–12 Discussion of the award panel during the "New Trade Fair Center Stuttgart" competition

minary check being capable of supplying correct and clear basic data for the decision.

2.1.9 Idea Finding through Investor Competitions

During so – called investor competitions, a given plot is generally written out for planning by either a community or another plot owner. The investors must supply development options as well as marketing and financing concepts (one example for this are the so-called Friedrichstadt-passagen in Berlin). The award is then granted to the best urban planning solution in conjunction with an offer of purchase for the plot or a respective partner deal.

There is a different situation, however, for the investor competition according to PGC that is currently under investigation: here, there is a condition that the respective investor can only obtain the plot if there is an agreement that the award-winning competition design is then actually also realized.

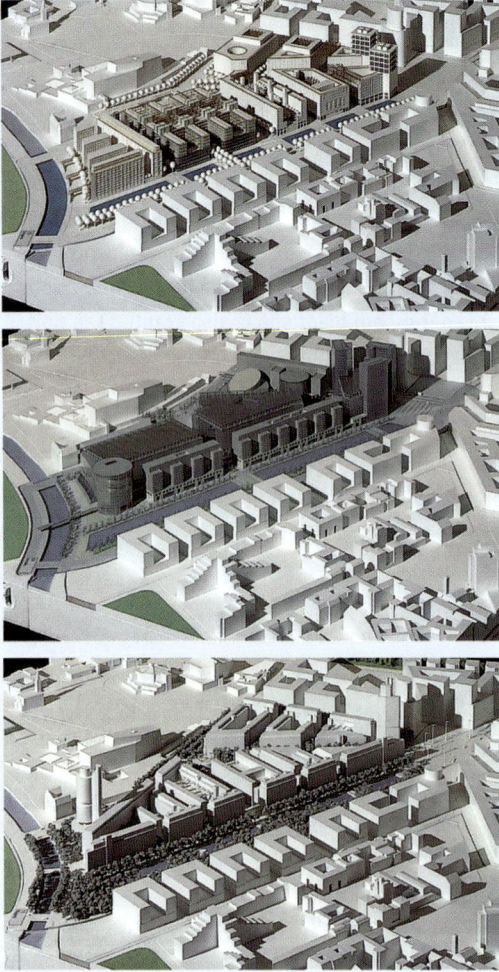

Fig. 2–10 Alternative designs (Potsdamer Platz in Berlin)

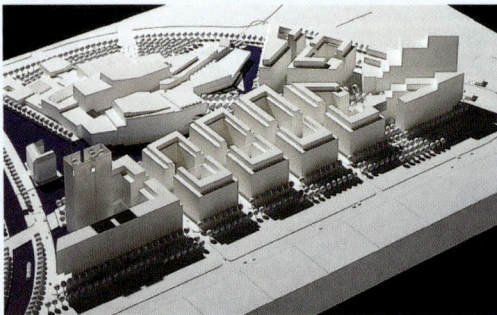

Fig. 2–11 Winning design by Renzo Piano (Potsdamer Platz in Berlin)

2.2 From Preliminary Planning to Award of Contracts

2.2.1 Preliminary Planning

During the preliminary planning stage, the following planning aspects usually need to be clarified:

– Functional context (spaces, traffic routes)
– Building geometry (cubic capacities, fundamental facade design)
– Energetic system (building physics-related frame data, building engineering),
– Constructive system (building grids, construction grids, floor heights)

Following conclusion of the ideas stage, it is now time to instruct the relevant specialist planners and surveyors. The architect works out the preliminary planning at a scale of 1:200, where all specifications concerning building geometry and facade ought to be decided on. In doing so, he receives advice from specialist planners and surveyors, e.g. in regard to construction grid, building heights and facade formation.

It is advisable, on the basis of this preliminary planning, to undertake an energetic system simulation in order to obtain secure planning specifications for further execution. To this end, the entire geometry of the building, building physics data and weather conditions as well as intended use of the building are simulated via computer in order

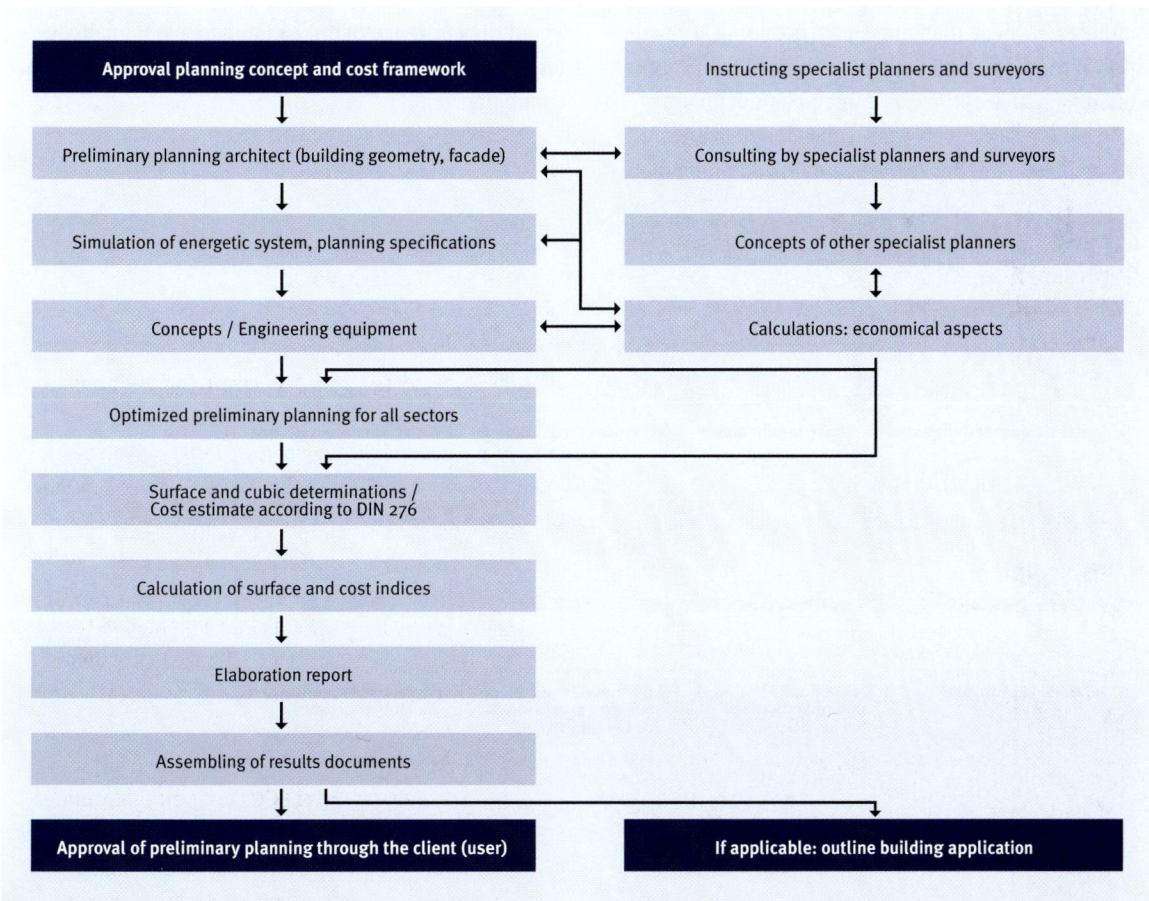

Fig. 2–13 Preliminary planning stage

to allow for room conditions to be calculated for the various conditions. This allows for creating an optimum building envelope in respect to intended use and planning by the architect, the date of which is to serve as basis for further planning.

On the base of this simulation and the available plans, the specialist planners create their concepts for the engineering equipment. These relate to the specifications for the architectural competition and incorporate aspects of later operation (Facility management). To this end, user requirements are converted into process requirements and finally into an administration concept.

Hence, project management ensures that technical planning is optimally adjusted right from the start to the aspects of later operation.

Parallel to all these planning steps, economical calculations are going on all the time with the aim of creating a building that is optimally designed in functional, quality and economical terms. Following delivery of the complete planning services, optimized preliminary

planning for all specialist areas is shown, whereas the concept for outdoor and outside areas also needs to be included. Based on the available plans and calculations, required areas and cubic content are determined as well as a cost estimate according to the breakdown of DIN 276 (costs for structural engineering) undertaken.

Based on all available plans, data and calculations, an explanatory report is then created, which should also include the essential area and cost indices.

The collected documents are then submitted to the client for approval and, if applicable under inclusion of inspection remarks, signed for release.

At this point, it may be advisable for many projects to submit a so-called outline building application, meaning to notify the respective approval authorities concerned about the existence of the preliminary planning document in order to obtain statements as to its permission capability.

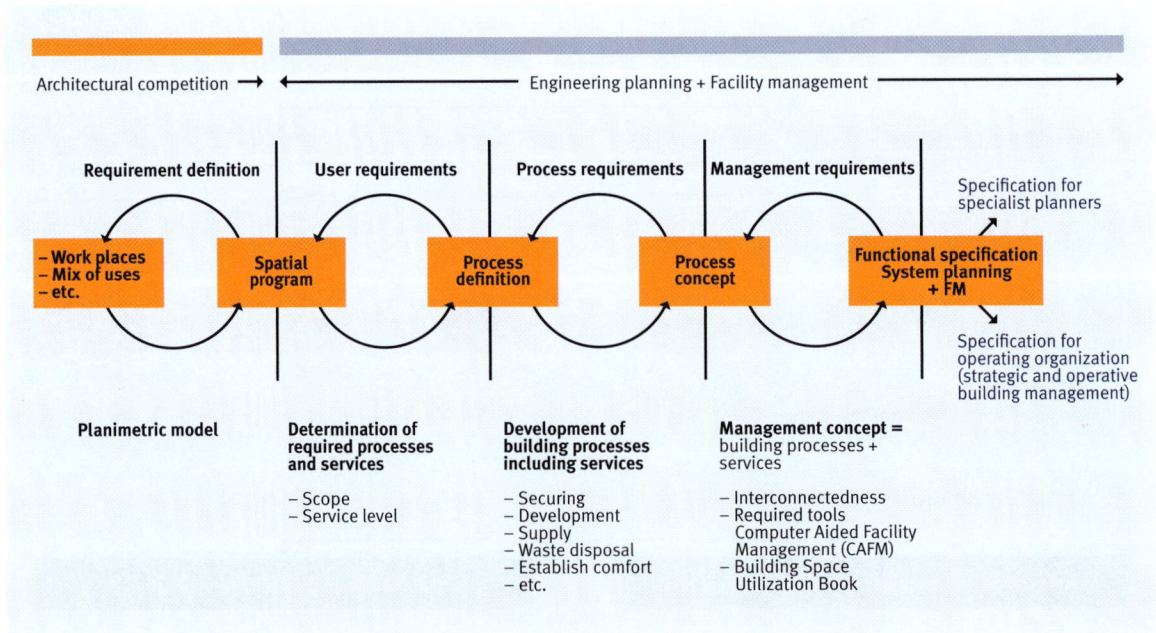

Fig. 2–14 Specification for specialist planners and operating organization

2.2.2 Schematic Design and Construction Design

In the schematic design phase, constructive details as well as principal ideas for roof, facade and fitout are to be developed.

Initially, the architect creates a design concept at a scale of 1:100, whereby development of facade and roof details are to be executed in parallel. For larger projects with a complex facade, it is advisable at this point to involve a specialist facade planner.

On the basis of the design concept by the architect, the specialist planners work out their projects for building engineering equipment for the individual trades. These projects are to be worked out with a kind of completeness and accuracy that allows for a call for bids for the individual trades. On the basis of the design concept, the structural engineer executes measurement of the individual components as well as their constructive formation. As a final result, we then have position plans and tender documents, which nearly conclude the service provision of structural engineer for this stage.

When the developments of the specialist planners are available, the architect needs to integrate these into his/her design and, when doing so, analyze all planning for their conformity and/or incompatibility. An important task of project management is in securely monitoring this service prior to transfer construction planning. Afterwards, relevant details for the construction permission need to be checked into also.

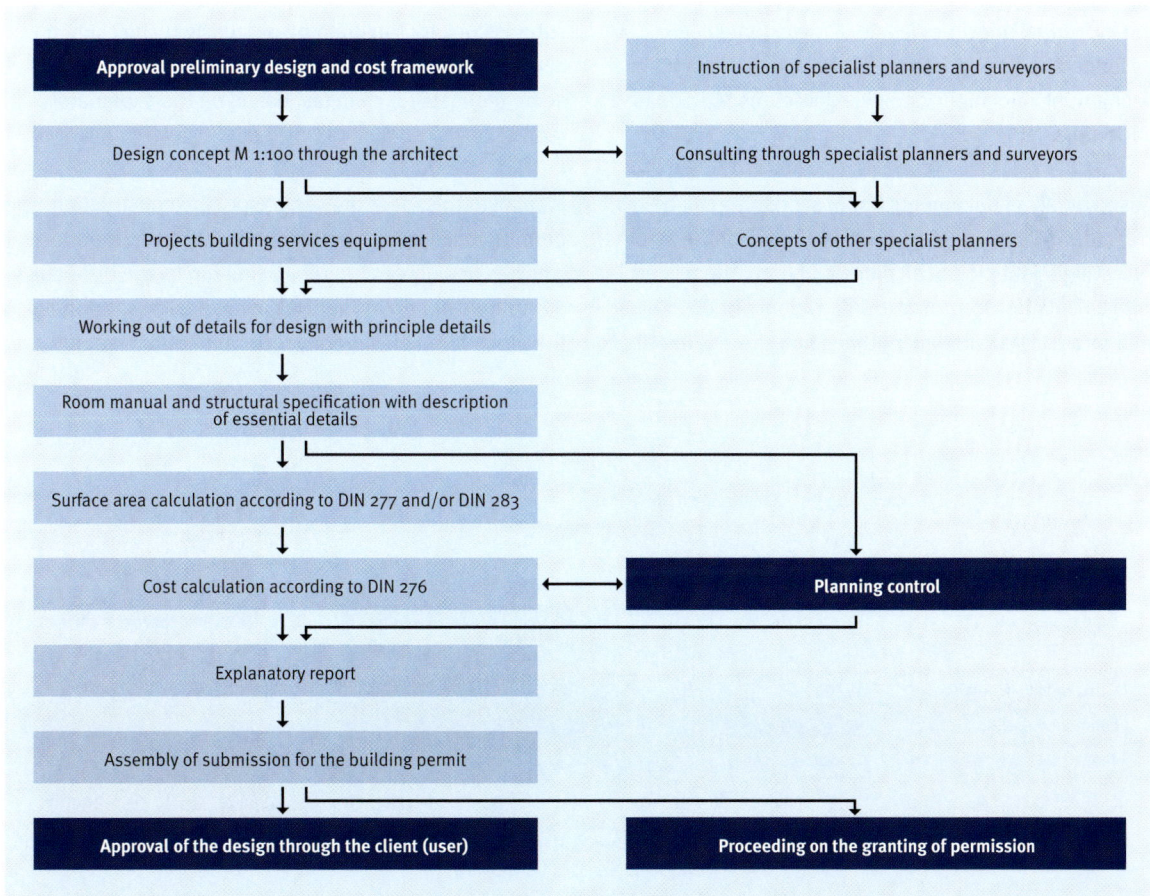

Fig. 2–15 Design stage

Fig. 2–16 Point by point coordination of planning details

The schematic design result is then sensibly outlined in a building space utilization book and structural specification including the essential details.

This document can be designed in such a manner that the individual principle details of the extension are described on individual sheets and, at the same time, segmenting into minimized overview plans M 1:100 is undertaken.

Based on these documents, room schedules can be created, which can later be used as a basis for the tenders. Following stipulation of all plans and data, there is an area calculation according to DIN 277 and/or DIN 283 as well as a cost calculation. At this point, there is another check into whether the cost framework specified can be adhered to with the now fixed principle details or whether there are any larger scope deviations. This knowledge is then put together in an explanatory report, which is handed over to the client together with the submissions for the building permission. The client now also grants approval of the design but this may include some requests for changes, based on the explanatory report.

Parallel to this, or at the same time, the construction permit procedure can be initiated with the authorities, which is usually executed by being run by the different authorities or public agencies involved. For shortening the procedure, it is urgently recommended to supply

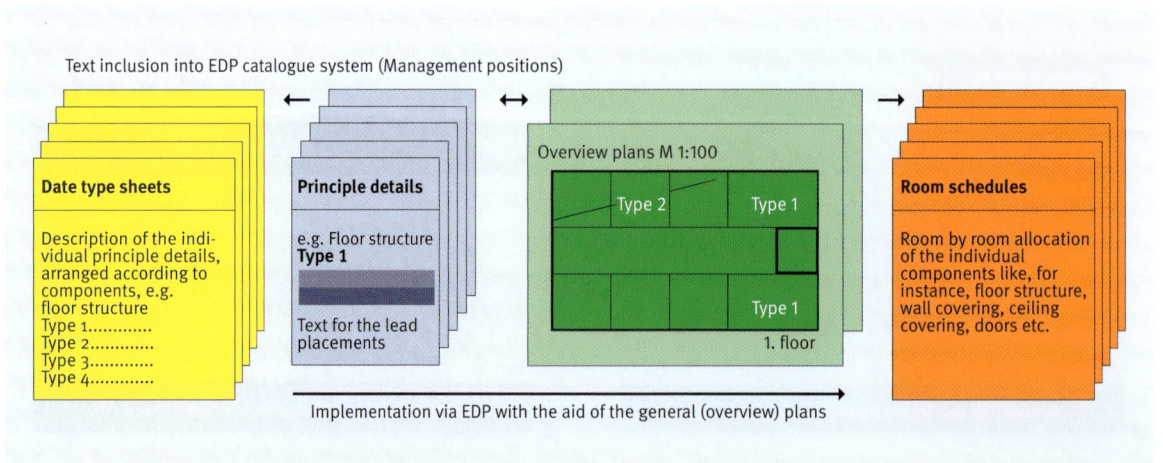

Fig. 2–17 Building space utilization book and structural specification as the basis for room outline

all sites involved with a copy of the permission documents and to personally make sure that these are also
being processed.

2.2.3 Models and Simulation

For the purposes of justifying both preliminary design
and design to client and permit authorities, the creation
of building models at an according scale is an important
aid. For clients coming from a different field, especially,
it often applies that they cannot get a clear idea of
planning content when they only have the planning
documents available.

Using the aids of CAD that we have available nowadays,
it is advisable during the design stage to supplement the
models with computer visualizations.

Visualization can also extend as far as depicting interior
spaces, especially when it comes to marketing prior to
construction being completed.

Using such imaging, potential buyers can, early on,
gain a better idea of the property offered than they could
if they only had drawings available.

Fig. 2–18 Wood model Potsdamer Platz in Berlin

Fig. 2–19 Building visualization of Potsdamer Platz in Berlin

Fig. 2–20 Interior space situation in residential units at Hohenzollernpark

2.3 Permission Procedures

Depending on the location of the plot, a project is subject to more or less complex permission procedures. These, independent of the other tasks, require additional time effort, which needs to be carefully logged, scheduled and coordinated by project management. In essence, this is a three-stage process.

– Development plan (binding land use planning)
– Building permission procedure (project-related planning permission)
– Construction approval (approval for building construction)

We will now look at the various steps.

2.3.1 Development Plan (binding land use plan)

Based on the land use plans, development plans are created for the building areas or the specific land-use areas outlined. Their applicable time frame spans approx. five years. They point to the permitted development options for these land-use areas, which, as a rule, is defined by the following specifications:

– Manner of constructional use: e.g. Residential development or mixed-use development
– Local traffic areas: these are, as a rule, access roads
– Scope of constructional use
  – Floor Space Index (FSI):
    Plot area x FSI = Total area of permissible floor space from ground floor
  – Cubic Index (CI):
    Plot area x CI = max. Cubic content above ground
  – Number of entire floors (NF): e.g. IV = 4 entire floors
– Plot area to be built on
  Site occupancy index (SOI): Plot area x SOI = max. Plot area to be built on

According to the new Federal Land Utilization Ordinance (BauNVO), underground components are also to be counted for plot area to be built on. This is supposed to reduce sealing of the soil surface.

As a rule, larger development plans are nowadays executed on the base of a policy draft supplied by the client since, correctly, one has arrived at the conclusion that it is not so much specification of numbers but, rather, working out the planning process which is to be taken as a guideline for correct development.

In the course of either creating or amending development plans, it is inevitable that one needs to work closely together with the authorities concerned. For the purpose of speeding up the creation of the respective plans, assistance can be offered especially in the area of measuring services. What is decisive, however, is to carefully work out the required steps for execution of the development plan, to equip them with realistic deadlines and then incorporate them into the overall schedule. As a rule, we need to take into account a total duration of ca. 1,5 years, at the very least however of one year. Complex development plans, involving massive public or neighborhood interests, may take significantly longer.

Once agreed upon, the development plan must be approved by the regional council and then forms the basis for the construction permission.

The graph shows a simplified sample process.

2.3.2 Building Permission Procedures

The building permission plans need to be created on the basis of the available development plan. The building authority and the other permission-granting authorities, including public agencies and also adjoining owners will then inspect conformity with the development plan. Smaller deviations, here, are generally tolerated by the permission-granting authorities while adjoining owners tend to show less willingness for tolerance. If one plans not to adhere to an existing development plan to a greater extent, a development plan amendment procedure is usually the result. In principle, this means a complete new process of a development plan procedure.

Administration/Planners	Local council	Community involvement	

Plan preparation decision

| Draft plan preparation decision | → Plan preparation decision § 2 BauGB | → Announcement official journal | → Information public agencies § 4 BauGB |
| Urban planning general principle template BBet. | → Decree civic participation § 3 BauGB | → Announcement official journal | → Preponed civic participation § 3 BauGB |

Decision on public display

| B-Plan draft with justification | | | Coordination with public agencies |
| Submission and statement administration | → Display decision § 3 (2) BauGB | → Announcement official journal | → Public display |

Resolution to adopt

| Verification concerns and suggestions | → Decision about concerns and suggestions | | |
| Limited participation according to § 2a (7) | → Bylaw resolution § 10 BauGB | → Announcement official journal | |

Approval, Comming into force

| Submission for approval | → Continuation decision regarding concerns and suggestions | | |
| Announcement and approval procedure. § 11 BauGB | → Approval acceptance | → Announcement official journal | → Coming into force of the development plan |

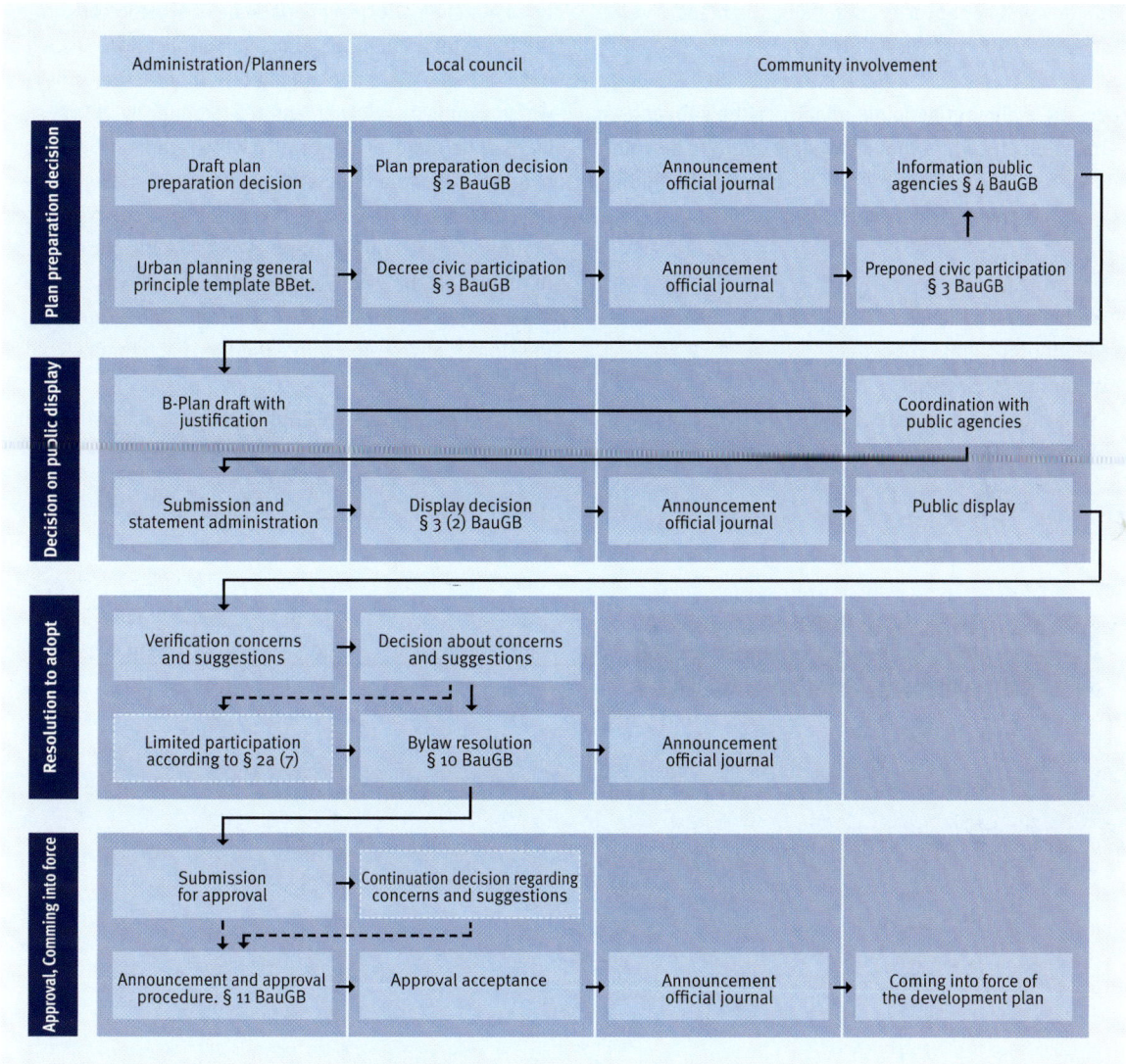

Fig. 2–21 Example for the process surrounding development plan procedure

In exceptional cases, an "accelerated procedure" according to § 33(2) BauGB can be undertaken instead of a development plan amendment procedure. In that case, the regional council approves a building permission granted by the building authority (without a corresponding development plan needing to be in place), as part of an exemption. The development plan amendment then follows. In practice, however, it has turned out that time gained through such a process is quite insignificant as a rule, since the building permission process itself tends to take much longer.

Generally, building permission procedures for saving time can be undertaken parallel to the creation of development plans. The risk in this case, however, is with the client who must assign the respective services at his or her own cost.

If the development process then fails, the respective money spent on costs is simply lost.

There are a number of points relating to the subsequent permission procedure that absolutely ought to be clarified with relevant authorities in the course of preliminary planning. These include, especially:

– Fire protection
– Work place ordinance
– Neighboring rights concerns
– Water rights concerns
– Excavation and removal options for excavation material, especially when it comes to contaminated soil

Usually, schematic design at a scale of M 1:100 constitutes the basis for a building permission application, since it is only for this scale that planning certainty can be achieved, which avoids a variety of textures. Frequently, however, large scale projects, especially, which work on significant overlapping of planning and realization, come with huge time pressure in their wake, meaning that a complemented preliminary planning M 1:200 is used for the construction request. This, however, is often connected with a huge amendment effort inclusive of frequent subsequent permissions, which come with extra fees each time.

The process of the permission procedure is illustrated in the following diagram, in a simplified manner.

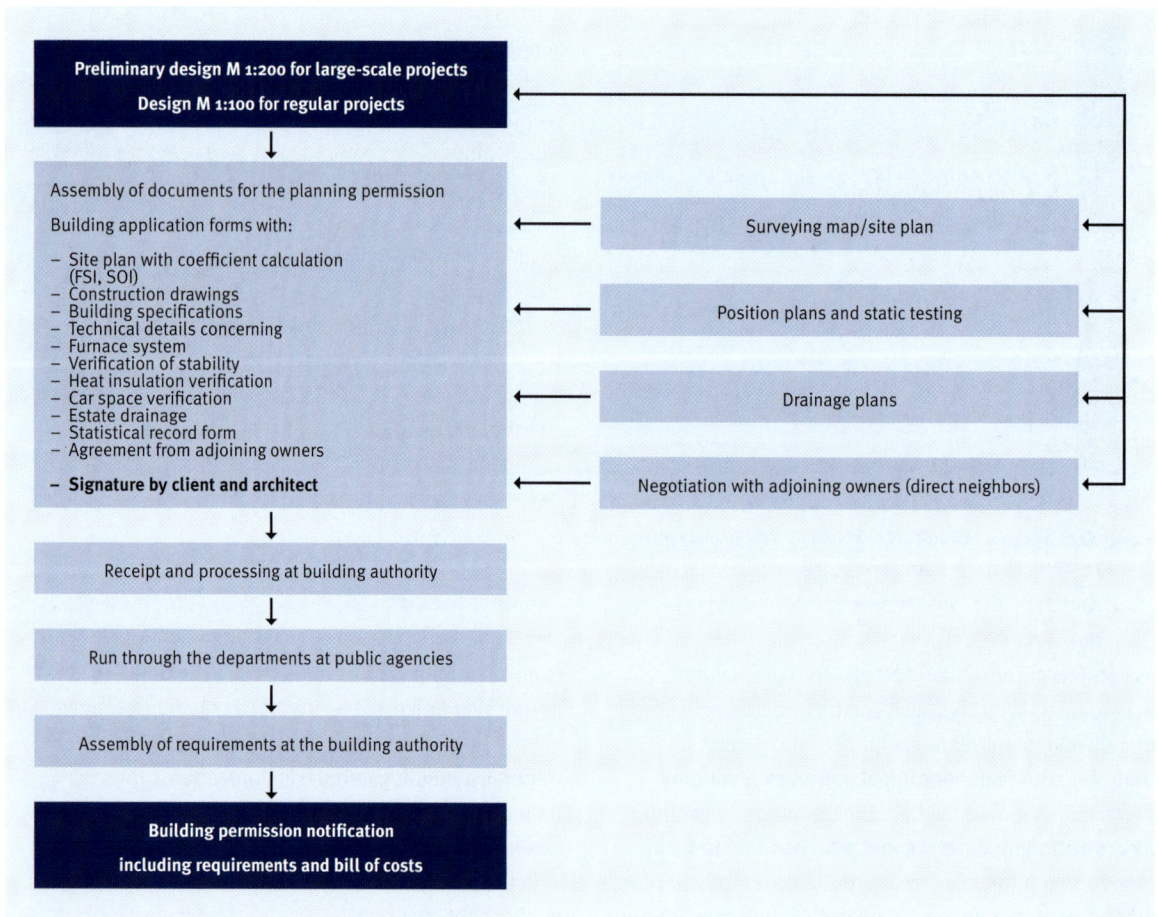

Fig. 2–22 Schedule process of a permission procedure

2.3.3 Building Approval

Building approval (the famous "Red point") principally is only granted once the final plans have been tested for stability criteria and drainage plans are also available. In the case of major construction undertakings it is generally a matter of obtaining planning approval by stages as only few verified plans are present at the commencement of the works. In the interest of the client, however, the project manager needs to take care that prior to commencement of the foundation works all position plans for a construction segment as well as the entire stability criteria have been verified. When neglecting this requirement, there can suddenly be significant additional costs for the client, especially if the structural engineer suddenly needs to act on plan amendments to undertake constructional changes in the upper floors down to the foundation (this has all already happened in the past!).

2.4 Execution Planning

Nowadays, two fundamental planning procedures are applied for the creation of planning documents:

– CAD 2D
– CAD 2D/3D

Today's CAD processes are usually two-dimensional but are increasingly being replaced by three-dimensional components. The advantage of CAD planning is in data transfer between the different planners and in joint plan updating. For large projects, CAD data is nowadays stored on an Internet server (also see chapter 4.7 Project Communications Management)

2.4.1 Execution Planning of Building Shell

Execution planning involves creation of plans ready for execution on the basis of an approved design for building shell, building services engineering, fitout, and outdoor facilities. For reasons of schedule, start of execution planning can often not wait until construction permits have been granted. Rather, this is done immediately after the client and/or the occupant have approved the design.

Basic building plan 1 (Ground plan, cross-section, view):
The so-called basic building plans are transformed, from the schematic designs that were approved by the client, via CAD into a larger scale, as a rule this is M 1:50. Dimensions entered are the main dimensions outside and inside (no dimensional chains), modular dimensions and dimensions from statistical preliminary computation. Only constructionally provisioned openings are entered as far as openings go, like, for instance, installation shafts, elevator shafts and large wall penetrations. Very important is sufficient elaboration of the cuts.

Formwork plan 1 and calculation of reinforcement plans:
On the basis of the basic building plan 1, the structural engineer then draws the formwork plan 1 M 1:50. This includes any and all structural members made of cast-in-place concrete, brickwork, prefabricated components, steel components, wood components and plastic parts and as much as they form part of the supporting structure or are constructively associated with the same (e.g. no walls as brickwork that have a solely dividing function but certainly walls made from cast-in-place concrete and parapet slabs from concrete prefabricated components).

Parallel to formwork plan 1, final calculation of reinforcement plans and precise dimensioning of structural components is undertaken and the resulting dimensions are then entered into formwork plan 1 (no dimensional chains, however).

Building shell design: Supplementary to the basic building plan, the architect develops all those details that are relevant to the building shell. For this, individual depictions at a scale of M 1:20 to M 1:1 clearly illustrate all essential material, construction and connection problems.
To be recorded here, especially, are:

– Connection points for facades, balustrades, balconies, ceilings, supports
– Roof structures
– Anchor plates, anchor rails

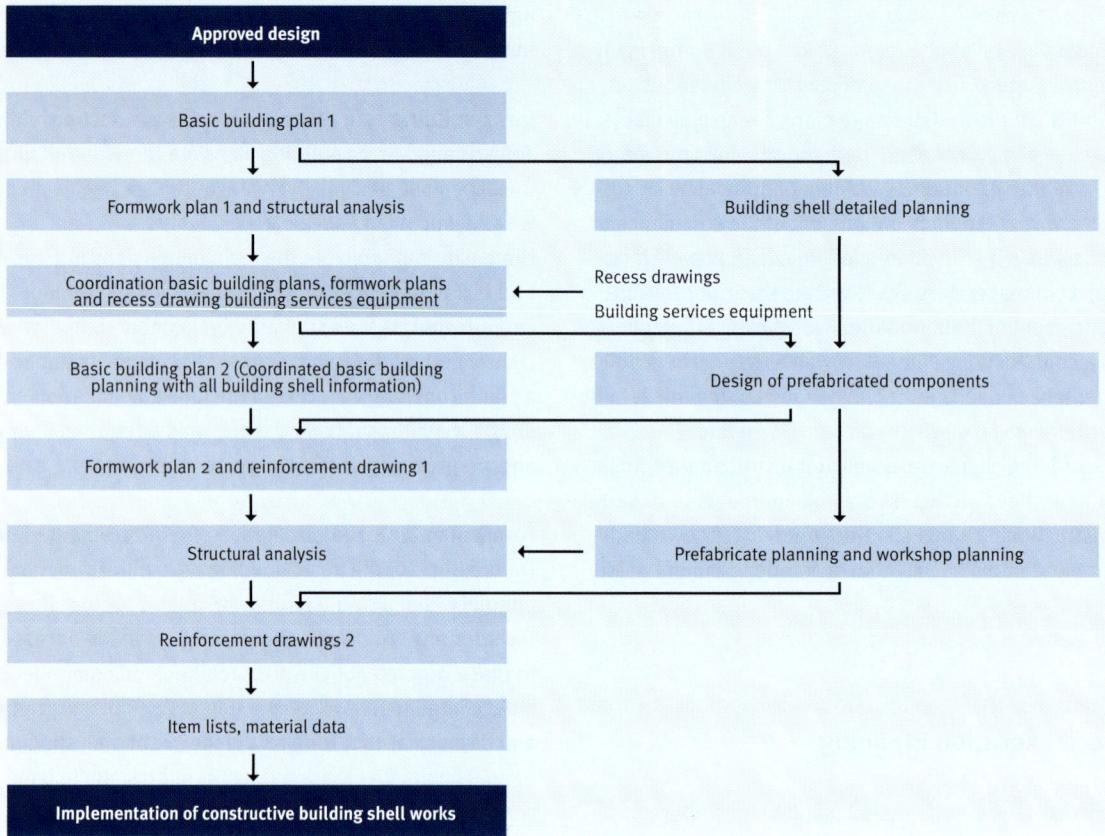

Fig. 2–23 Scheme process execution planning shell work

Recess drawings building services equipment: Based on their projects, the engineering experts enter their ideas regarding recesses, mounting openings, floor, ceiling and wall ducts, installation scaffolding, mounting rails and foundations. Further, envisioned dimensions for all larger equipment are to be specified, if possible on a specialized central plan M 1:20. If the schedule allows, the formwork plan ought to be used as a basis here in order to allow for arrangement of recesses in a manner that does not affect critical points. Any and all information is to be entered by way of circulations and sensibly the following sequence is usually selected:

Sewage pipes ⋯﹥ Ventilation ⋯﹥ Sprinkler ⋯﹥ Water ⋯﹥ Heating ⋯﹥ Electric lines

Coordination: The basic building plan 1, formwork plan 1 and the slot plan are verified and coordinated by the architect in conjunction with the static engineer and the engineering experts.

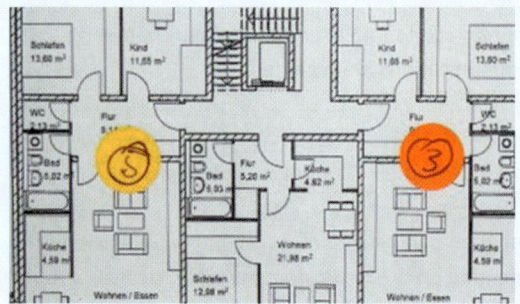

Fig. 2–24 Coordination basic building plans

In those areas where the various ideas are not in conformity with each other, compromise solutions need to be found. In this, construction should be given preference, if possible, since overall extent of changes is almost always significantly larger than was initially assumed.

Basic building plan 2: In basic building plan 2, coordinated information of the structural engineer and the engineering experts is adopted and dimensioned (dimensional chains, building shell). In this state, the basic building plans go to the structural engineer together with the building shell details and the prefabricated component plans. This basic building plan stage forms the basis for execution of the installation of non-supporting masonry. Prior to delivery of basic building plan 2 to the project development company, all brickwork and balustrade heights, possibly reinforcements and formations also materials involved, need to be specified. Basic building plan 2 needs to be identified with the remark "approved for the execution of any and all building shell works".

Prefabricated component plan: If certain building components are of prefabricated load-bearing elements, corresponding plans are to be created following coordination, parallel to basic building plan 2 (building shell area), while taking into account building shell details. Depiction ought to be at the scales of 1 : 10 and 1 : 20, steady correspondence needs to be established with work plan 2. This concerns mainly building components of reinforced steel, steel, wood and composite elements. These plans must be available for the creation of the formwork plan 2 and serve as basis for invitation to tender and production planning by construction companies.

Formwork plan 2 and reinforcement drawings: Formwork plan 2 includes all results from basic building plan 2, building shell details and prefabricated component plans. Any and all recesses, structural components, anchor plates, halfen rails etc. need to be registered. All depictions are now supplied with dimensional chains. Formwork plan 2 serves as supporting document for reinforcement drawing 1. Further, it is released to the building site for formwork preparation during workshop planning of the prefabricated components.

Reinforcement that becomes necessary after the design engineer then enters structural analysis into formwork plan 2 (possibly into a master print). Care needs to be primarily taken that the reinforcement is constructed in an assembly-friendly manner (absence of over-length, few bending up areas, as much as possible: storage mats rather than list mats due to long delivery times). The individual posts should have information concerning both length and diameter.

Prefabricated components – workshop planning: Based on the Bill Of Quantities, prefabricated component plans and formwork plan 2, the prefabricated component manufacturer draws the workshop plans, For reinforced concrete prefabricated components, reinforcement is mainly calculated according to manufacturer tables and then entered, as well as all the anchoring points. Since the undertaking also requires installation of structural components of all kinds, the different structural components (type, manufacturer) as well as delivery need to be agreed upon prior to commencement of planning.

Structural analysis: According to the implementation provisions of the German Building Test Ordinance (Bau-Prüf VO), construction may only commence once verified structural analyses and verified construction plans are available. For very simple buildings, building authorities handle the verification, for more difficult buildings, test institutes or freelance test engineers are asked to become involved. The following documents must be made available for the testing process:

– Structural analysis (Determination of cutting forces and dimensioning)
– Formwork plan 2
– Reinforcement drawing 1
– Execution plans prefabricated components
– Calculations and details concerning facade anchorages
– In case of excavated trench shoring: trench lining statics and trench lining plans (cuts, executions)

Following inspection, the documents – now equipped with inspection remarks – are sent back to the plan creators.

Bill of material and material data: The verified documents are corrected according to the inspection remarks in green and possible disagreements are dealt with.

From the corrected plans, posts and individual components are then extracted and compiled into bills of material, e.g.:

- Steel lists for reinforcement
- Item list for cans, empty conduits in concrete prefabricated components
- Item lists for individual parts for welded steel girders
- Item lists for girders, anchors and construction wood at Berliner Verbau

The corrected documents and the item lists are copied in the numbers required and then released for execution (building site).

Special care must be taken that any and all documents are only released via construction management on the side of the client and that they maintain a precise plan release log. This document must also include all changed plans, complete with release date and change index, without which they must not under any circumstances be sent on to the building site.

2.4.2 Execution Planning of Building Services Equipment

Execution planning is undertaken on the basis of position plans and basic building plans of the architects.

Execution planning H, V, S, E (Heating, Ventilation, Sanitary, Electrics). Execution plans for building services equipment can be either created by the engineering experts or by the construction companies.

Execution plans for the remaining trades of building services equipment are created on a scale of 1:50 on the basis of the projects and of the basic building plan 2. The plans must include the following specifications:

- Ventilation: Ducts (routes, dimensions, branching), mixed boxes, outlets, anemostats, equipment erection
- Heating: Pipelines (routes, dimensions, material), radiators, connections, valves, shut-off devices, pumps, ventilation ducts
- Sanitary: Pipelines (routes, dimensions, material, connections, outlet points etc.)
- High and low voltage: Main distribution, storey distribution, distribution boxes (cross section, type of cable, cable racks etc.), wiring diagrams, consumption outlets

Execution plans serve for ordering of materials and often directly for installation; sometimes, separate execution plans are created for ventilation. Further, they form the basis for the workshop plans.

Execution planning conveyor systems: For conveyor systems (elevators, escalators), there are no execution plans per se; here, workshop plans and assembly plans are directly created by the firms, since these systems are very manufacturer-specific.

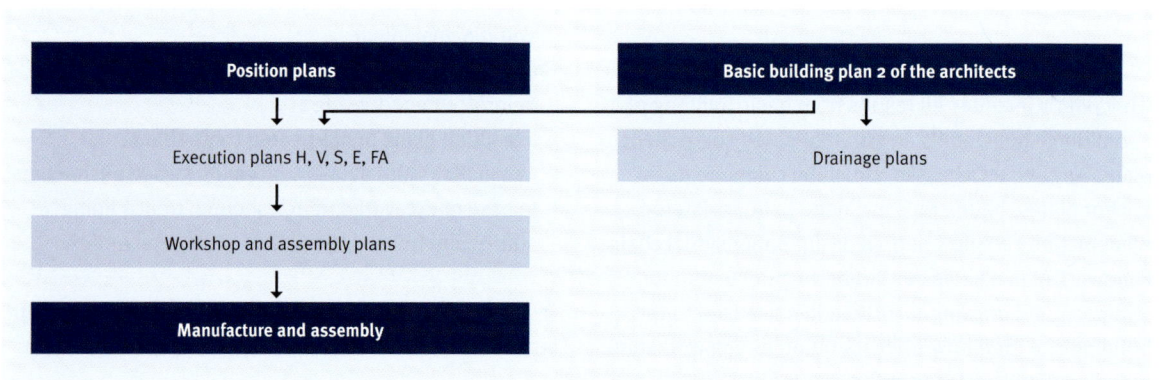

Fig. 2–25 Scheme process execution planning of building services equipment

These trades are to be assigned prior to the building shell work or at least simultaneously to it in order to receive building shell specifications in time and remain in the framework of procedure AF 5 (slot plan).

Workshop plans: The workshop plans contain specifications for manufacture and pre-assembly in the factory. The assembly plans contain all specifications for assembly and installation at the building site. For example, these are some of the workshop plans to be created:

– Ventilation: for manufacture of ducts, equipment and control cabinets
– Heating: possibly for radiator manufacture (not for radiator produced in series)
– Sanitary: For pre-assembly of cells, installation blocks and manual pre-assembly
– Electric: For the manufacture of distributor cabinets, phone switchboards, equipment

Drainage plans: Sensibly, drainage is incorporated into the foundations' formwork plan, where generally formwork plan 1 suffices. In case of drainage pipes being located in the influential region of the foundations – especially when it comes to large building facilities or grid-like foundations – fine-tuning with the foundation plan is required (bridging, encasing in concrete of the pipes etc.). Specifically, the drainage plan needs to contain:

– Routes, gradient
– Dimensions, material
– Connections, branching components
– Ducts (for revision, cleaning or branching)

Following coordination, item lists (pipes, branches) are extracted from the drainage plans; these entire document batches are then handed to the building site for the purposes of ordering and execution.

2.4.3 Execution Planning of Fitting Out

The basis for planning of fitting out is the basic building plan 2 of the architects as well as the room manual and the construction book.

Extension overview: Initially, an overview is to be created, concerning the fitting out details that need to be worked on and a decision is to be made concerning the extent to which prefabricated products can or should be used. The basic building plan 2 – after it has been equipped with all shell work specifications and approved – contains the possible specifications for a scale of 1:50 concerning fitting out. This includes:

– Floor structure
– Door and window hinges
– Wall, supports and ceiling claddings
– Roof structure
– Facade structure

In the M 1:50 views, coordinated depictions of visible components are entered, complete with obligatory specifications for mass determination and for application on site. These are depicted, among others:

– Facade components, maintenance balconies, cleaning facilities
– Non-supporting, prefabricated components
– Canopies, solar protection
– Superstructures, attics
– Advertising and marketing devices, lighting etc.

The views must contain additional information concerning related detailed drawings material specifications and if possible also color specifications. Once it is completed, basic building plan 3 serves as documentation for the call for tender of the extension, detail processing and also for the workshop and assembly plans of the executing companies. Further, it is used in an interim stage for the creation of the execution plans for building services equipment. For this, sanitary installations and heating radiator allocation principles as well as ceiling structure need already be clarified.

Detailed drawings fitting out: General indicators outlined in the overview of fitting out are now worked out at a scale of 1:10 to 1:1 and solutions are illustrated in individual depictions.

We distinguish here between two kinds of details:

- Individual details: This includes all aspects that do not occur more than once, e.g. floor structure in special rooms, details in the entrance area, details for safe installations or other security facilities.
- Regular details: This includes details occurring at great frequency – possibly with slight differences – over and over. This especially includes doors, frames, windows, solar protection, dividing walls, lights, wardrobes or cabinets.

Regular details are summed up in overview plans and then assembled into lists, that contain the various elements, specifying item numbers – at basically the same structure allowing for deviations from the basic element. Deviations can include:

- Dimensions
- Stop direction
- Opening movement, drives
- Infill material (Windows, dividing wall panelling)
- Color

These details, additionally to basic building plan 3, serve for execution and creation of workshop drawings. In the best-case scenario, they ought to be completed prior to the invitation to tender. Regular details and items lists may be used directly for ordering, in prefabricated components.

Detail coordination (Planners, manufacturers): Once an assignment has been placed, details specified by the planners must be coordinated with the manufacturers for production-related problems and use of company-specific elements.

Workshop and assembly plans fitting out (executing firm): Once details are coordinated, executing firms initially create workshop plans for production of the components and pre-assembly in the factory. Afterwards, assembly plans for installation and assembly on-site are created. They are submitted to the architect for approval to ensure production according to planning.

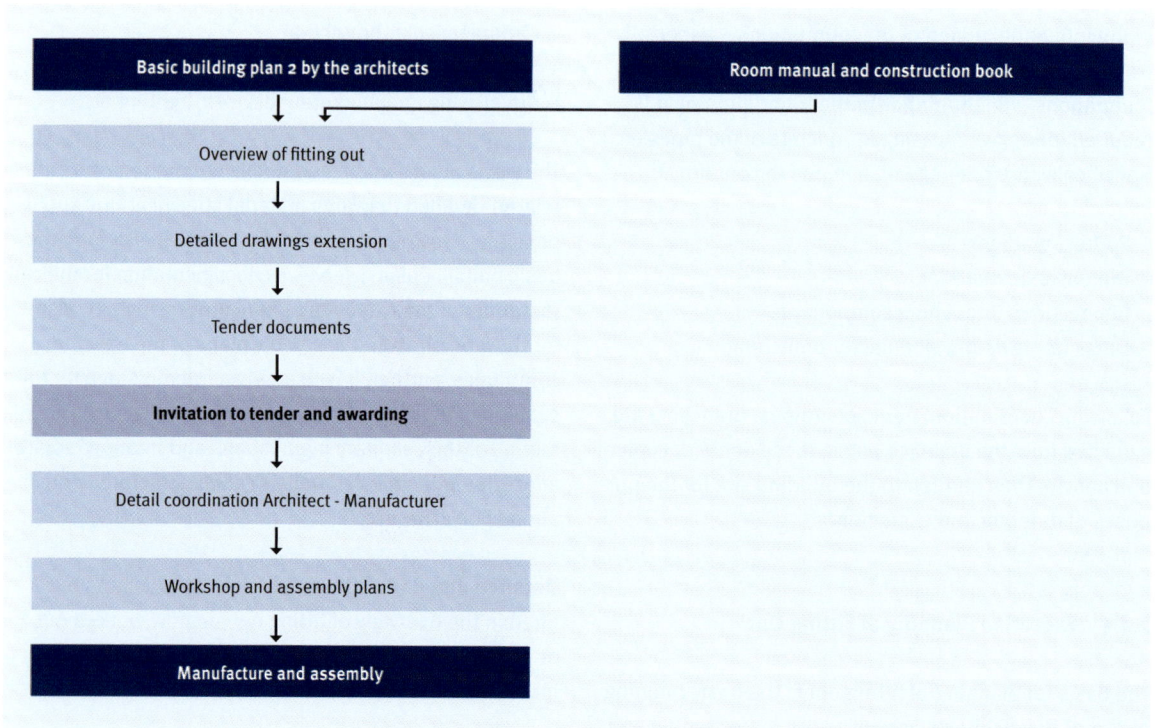

Fig. 2–26 Scheme process execution planning fitting out

2.5 Procurement of Construction Services

2.5.1 Conventional Tendering and Contract Awarding after FSAE

Parallel to the implementation documents, preparation for tender is undertaken. When there are tight deadlines, one does not wait for completion of the permission procedure but the invitation to tender commences already while the construction law procedure is going on. Prior to awarding, it is important that the construction permit is in place and that stipulations contained in the permit are considered for the construction contracts. If there is partial permission already, it can be called for separately and awarded, although this cannot always be recommended due to the problem of separating excavation, downpipes and foundation. The following process scheme can be considered as a basis:

Planning-related documents: Required planning documents are distinguished by the three main tender packages, whereby as a rule the most secure document would be the respective execution plans; however, waiting for these often means losing a lot of time.

- Package 1 – Building shell and conveyor systems: Here, schematic designs will suffice M 1:100, whereby, for the shell, estimated mass needs to be specified by the structural engineer (uncertainty factor!).
- Package 2 – Facade and building services: Here, too, schematic designs and/or project plans serve as the basis for the invitation to tender, whereby the client needs to make decisions as to technical equipment and quality
- Package 3 – Fitting out and outdoor facilities: For this package, execution plans, as a basis for the invitation to tender, are to be created especially for those trades that require workshop production – like, for instance, dividing walls, wardrobe walls, ceiling cladding, visual design. Here, too, decisions by the client are crucial when it comes to establishing a certain quality standard.

Creation of the Bill Of Quantity (BOQ): Based on the planning documents, individual positions are itemized and corresponding masses determined. The BOQ is reviewed by its creator and possibly by inspection entities, then corrected. The BOQ is printed out following the corrections, whereas required preliminary remarks and conditions are added.

Public announcement: Parallel to itemizing positions, the official announcement text of the tender in the press is created. Some 4 to 6 weeks prior to mailing out BOQ, the announcement is published (Official journals, large daily newspapers), for large contracts, EU announcement is required. Addresses of applying firms are recorded and the numbers of BOQ to be provided are specified. For restricted call for tender, only firms meeting the performance specifications are considered.

Tender processing, submission: Applicants insert into the BOQ their unit prices (possibly divided by salary/fee and material) and total prices; often specifications are made concerning sliding salary clauses and flat-fee surcharges. Tender documents must be submitted by the submission deadline in a sealed envelope.

Participation in the awarding process: In the case of (possible) presence of the bidders, quotes are individually opened and the quotation amounts are read out and recorded.

Bids are factually reviewed and from a figures point of view, price comparisons are created where the most important unit prices are compared. Further, information needs to be gathered about capacities and financial background of the bidder. On the basis of this information, an awarding proposal is then made to the client.

Client and architect and/or engineering expert negotiate with the bidder or also with several of them, in order to specify the final scope of the assignment and also contractual details. After that, final decision about awarding is made. The bidder selected receives an official letter of assignment that outlines all fundamental aspects of the contract as well as the gross order sum. The bidder thus becomes the contractor/agent and one of the contract parties.

2.5.2 Services Specification by Elements

Alternatively to a services specification according to the individual trades together with individual positions on the basis of a comparatively detailed planning, it is recommended to resort to an element-based description of components. This is the case especially everywhere when knowledge of the construction companies or the executing firms is applied in the form of prefabricated or system products. This, for instance, applies to windows, metal facades, dividing walls and suspended ceilings, system flooring, screeds with floor heating etc. The various elements/components are defined only via requirements and simple item numbers like windows (with complete dimensions) or metal facade (m²/glass proportion in %/construction physics data) and one overall plan view.

Experience shows this type of services specification leading to more favourable prices and better details than individual listings by architect details.

2.5.3 Lead Times

Often, the required plan lead times for work commencement for the various trades on site, which can affect the entire process concept, are totally misjudged.

Example awarding of building shell works: If a conventional process is chosen – meaning reinforced concrete or in-situ concrete construction – the following process varieties can result:

A plan lead-up time (until start of construction) of about 6 months results for commencement of basic building plan stage 1, whereby the different viewpoints of the architect (looks from above onto the floor) and the structural engineer (looks up to the ceiling from below) need to be considered. Hence, the architect already needs to commence about 6 months prior to construction of a reinforced concrete ceiling with final design of the floor lying above.

If all execution plans are supplied by the planners, assignment of the reinforced steel works suffices at V2 prior to start of construction. If reinforcement drawings are supplied by the building shell company, assignment must be undertaken at the time of V1. If the contract goes to a general contractor who also supplies the basic building plan of the architect then assignment must be undertaken at the very latest by the time of V0.

Example awarding of ventilation and air conditioning: This trade comes with a required plan lead time of about seven months from execution planning, followed by workshop planning and assembly planning as well as duct production. Awarding must be at the very latest by V1 since workshop planning is up to the contractors. Nowadays, it is customary to assign execution planning to the contractors. In this case, awarding must be at the time of V0, prior to implementation plans.

Example awarding facade: For more complex finishing works like facade or built-in cupboard wall system according to the visions of the architects, lead time is significantly more. (See Figure planning lead time and awarding facade conventional.) The BOQ can only be put together on the basis of completed basic building planning and it is very time-consuming since many details must be drawn and described. There is usually a long coordination stage between contractor and architect to make design production-friendly. One needs to take into account run-up times between 11 months and up to one year.

A functional services specification can shorten the lead-time by at least three months (Figure planning lead time and awarding facade by elements with requirements). Later, there is shortening of production and assembly times because contractors are better able to exploit their available resources.

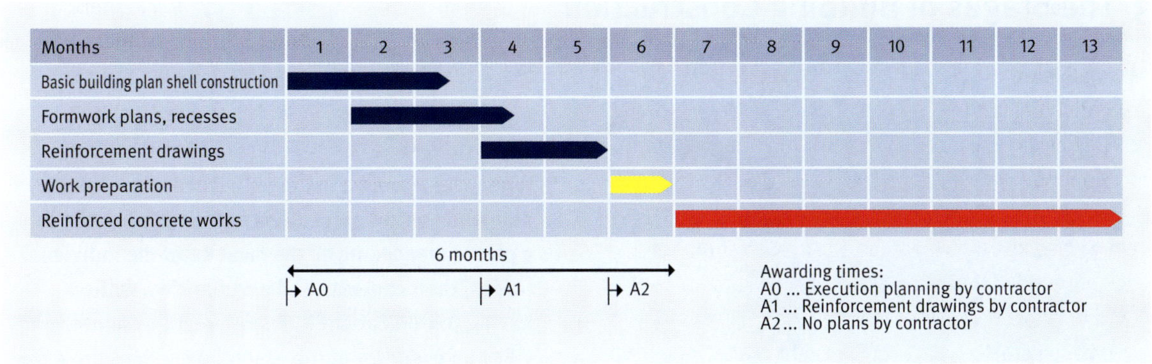

Fig. 2–27 Planning lead time and time of awarding of building shell works

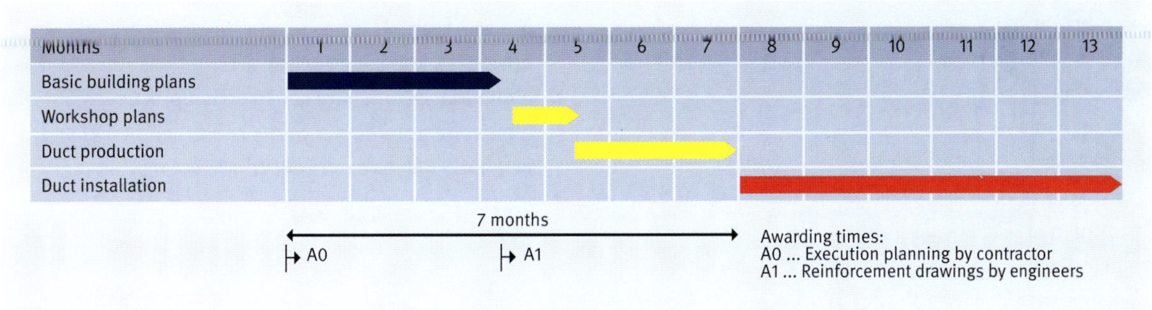

Fig. 2–28 Planning lead time and time of awarding of ventilation and air conditioning

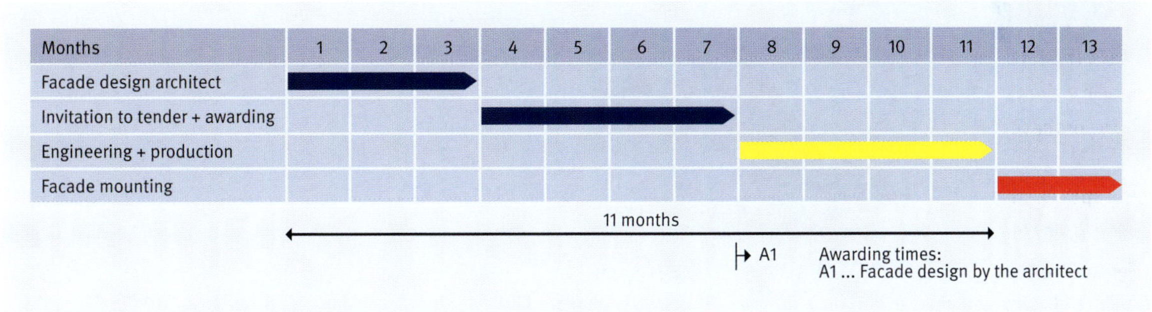

Abb. 2–29 Planning lead time and time of awarding of facade conventional

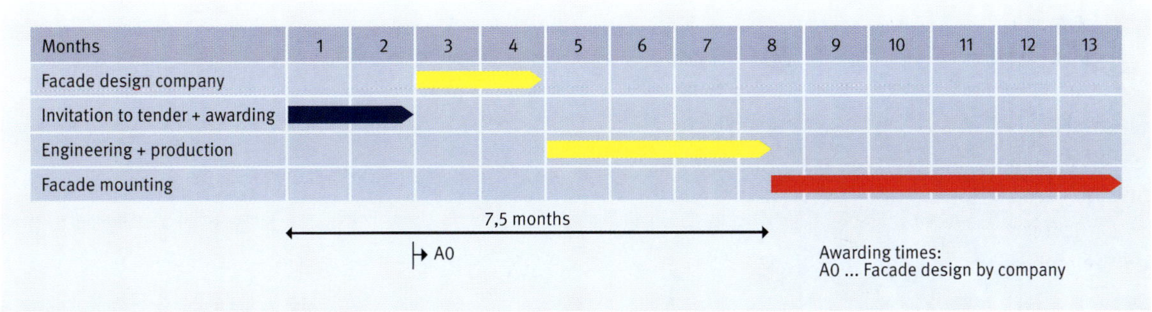

Fig. 2–30 Planning lead time and time of awarding of facade by elements

3 The Stages of Building Construction

Whenever a project manager wishes to coordinate, plan and control procedures at the building site in a proper manner, he or she must know the individual trades, their contexts and peculiarities, without having to think about it. Therefore, this chapter provides an overview of the essential services provided by the construction companies. The required depth of information here shown is the minimum required for process control. Here, we are looking more at principles and construction alternatives rather than individual details and construction processes but it must be pointed out that a construction manager generally delves even deeper into the matter.

3.1 Site Preparation and Site Facilities

Prior to commencing actual works, large construction projects initially require the creation of an infrastructure.

Coordination with the authorities: In order to run his/her site later on without problems, the site manager, during the construction preparation stage, must coordinate a huge number of individual points with the authorities involved. This includes, among others:

– Traffic regulation (possibly also rail way traffic)
– Traffic stress through construction site vehicles
– Road shut-off
– Road construction and pipeline construction
  by the authorities

This ought not to be left up to the shell work contractor alone, since too much time could be lost by this.

Construction site fence: The first task must always be securing the site with a suitable construction site fence (mesh wire, planks etc.) with stationary entry and exit gates in order to prevent unauthorized parties from entering the site and thus ward off connected accident dangers.

Preparation of the terrain: Once construction permission has been granted (red point) for demolition works, demolition and removal of existing buildings, trees, refuse etc. commences. In case of trees or groups of trees existing on the site that are to be retained, there are to be secured with massive wood barriers against any accidental damage (tree protection).

Relaying of lines, rerouting of main lines: If that existing pipe plans or construction permission stipulations point to the fact that there are public or private supply or disposal lines on the site, required rerouting – or often also shutdown – of these lines is to be coordinated with the parties involved and rerouting to be undertaken.
In the case of large, spacious sites, it is recommendable to immediately also lay future main supply and disposal lines, all the way to the ducts close to the building, in order to prevent having to dig up the earth outside the building pit later on.

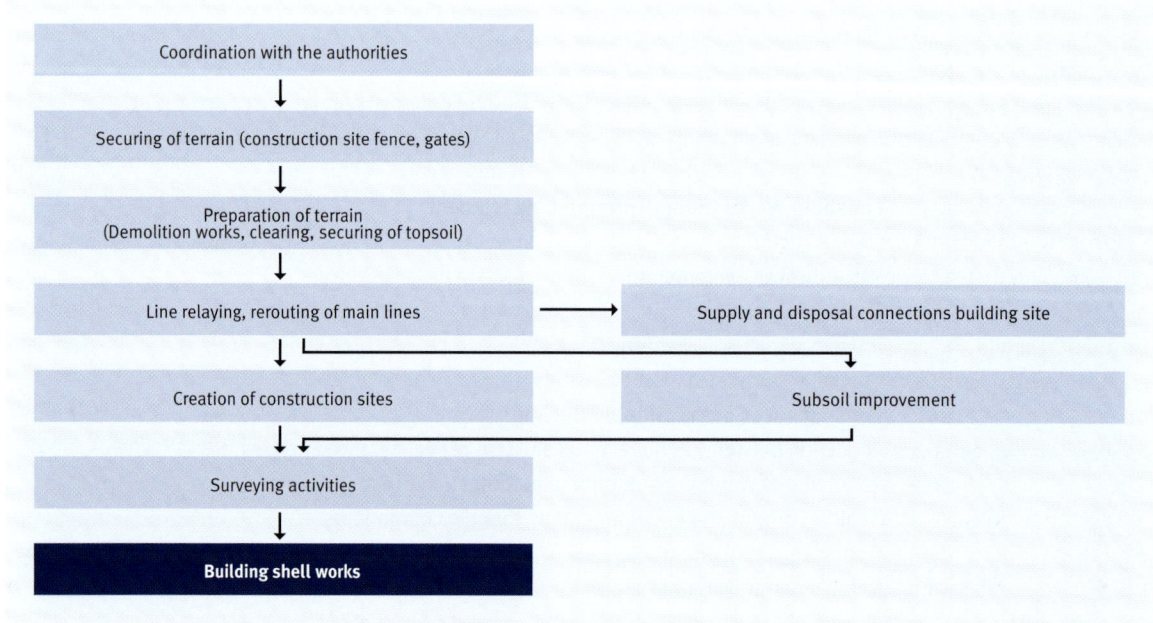

Fig. 3–1 Process: Preparation and development of the building site

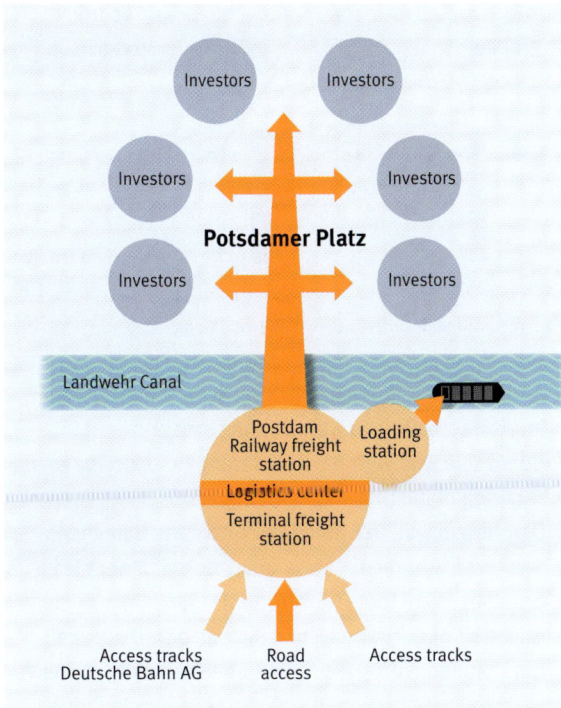

Fig. 3–2 Logistics concept for the Potsdamer Platz major project

Supply and disposal lines: To allow the contractor to commence with the works as soon as possible after awarding of the contract, it is recommended for large construction sites to create main connections straight away for construction site water and electricity supply and to make available sufficient space for storage areas and living options.

Construction roads: In the case of large building sites, cable laying is then followed by building construction roads. Construction roads that are only required for excavation and building shell works can be designed as gravel roads but the others, even those for transport of building services equipment and fitting out materials, are to be paved roads.

Subsoil improvement: If the soil report points to measures for being required improving the subsoil, this must be undertaken either at the same time or immediately after the construction roads.

In general, we distinguish between:
– Soil exchange
– Soil improvement
– Compaction of subsoil

If these methods turn out to be insufficient or too complicated, the pedestal to the base is possibly placed lower down or a special foundation type is chosen (see building shell works).

Surveying: Measuring of main survey points (reference points) as the basis for measuring the building through the contractor.

For very large projects, the construction site area reaches far beyond the actual plot. For the Potsdamer Platz project, for instance, external construction roads and bridges were built to access an external logistics center with rail and ship connection so as to not place extra stress on public transport networks.

Construction site facilities: This includes all facilities that the contractor requires for the provision of services, under adherence to the relevant regulations, for instance:

– Transport equipment (Cranes, concrete pumps, construction elevators)
– Mixing equipment with access routes or transfer containers for ready-mixed concrete
– Storage areas for casing, reinforcements, pre-fabricated components, bricks and other materials
– Barracks for construction management, accommodation, canteen and sanitary facilities
– Construction site electricity transformers, water hydrants

However, the required spaces are frequently not available for inner city construction sites; this must definitely be specified in the BOQ. Therefore, it is recommended that project management, prior to the invitation to tender, creates a construction site master plan that makes allowance for the possible infrastructure of the site. This site master plan is essential as the basis for tendering for construction site access, shell work and for creation of a realistic basic time schedule.

3.2 Shell Construction and Space Enclosure

The first parcel in construction is shell construction with space enclosure, meaning creation of the construction structure with a weatherproof envelope. As a rule, we can assume the following process for this:

3.2.1 Excavation Works and Construction Pit cladding Works

First, the topsoil needs to be removed and deposited on the areas provisioned for this, where it is stored. Then, actual excavation works commence according to the excavation plan, all the way to rough grading. Afterwards, smaller machines are used for fine grading and for excavation works for foundations and supply lines. If the pit cannot be scarped because it is either too deep or there is not enough space (downtown), the pit sides need to be secured through cladding. There are various options for this:

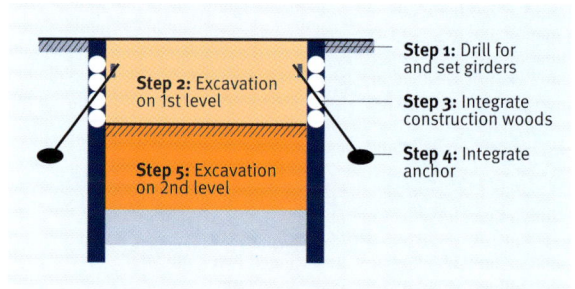

Fig. 3–4 Schematic process Berlin type pit lining

- Berlin type pit lining with back anchoring (for regular loads and buildings with working area)
- Bored pile wall with back anchoring (for extreme loads, securing of neighboring buildings)
- Diaphragm wall (for buildings where the lining is to serve at the same time as supporting wall or at least as protective wall for the insulation)
- Sheet piling with the same effect as bored pile wall but rammed (very expensive). Sheet piling is rarely used for construction pits in structural engineering.

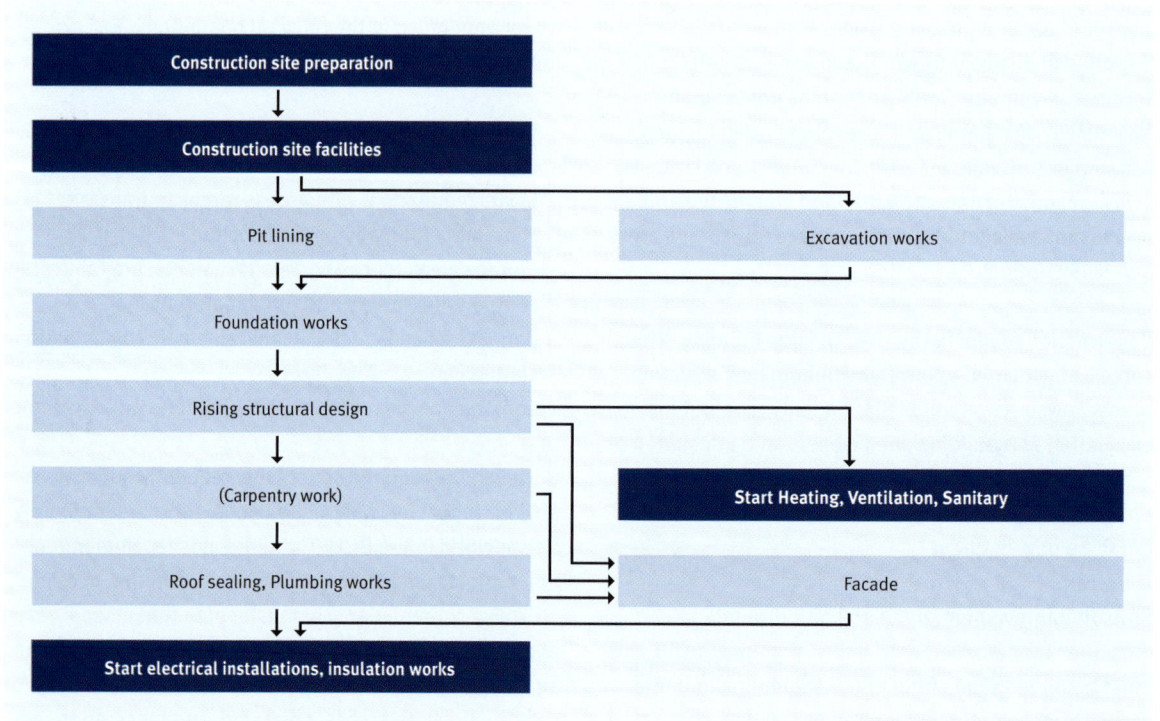

Fig. 3–3 Process building shell construction and outer skin

The scope of excavation work generally depends on the following parameters:

– Size of building pit (the larger, the more work is required)
– Depth of pit (the deeper, the less complex the works)
– Soil makeup
– Transport distance to depot and traffic situation

– Requirement for pit lining
– Type of pit lining
– Grading requirements

There are special requirements for ground water excavation works. Fig. 3–5 shows underwater excavation at the example of a major construction project in Berlin.

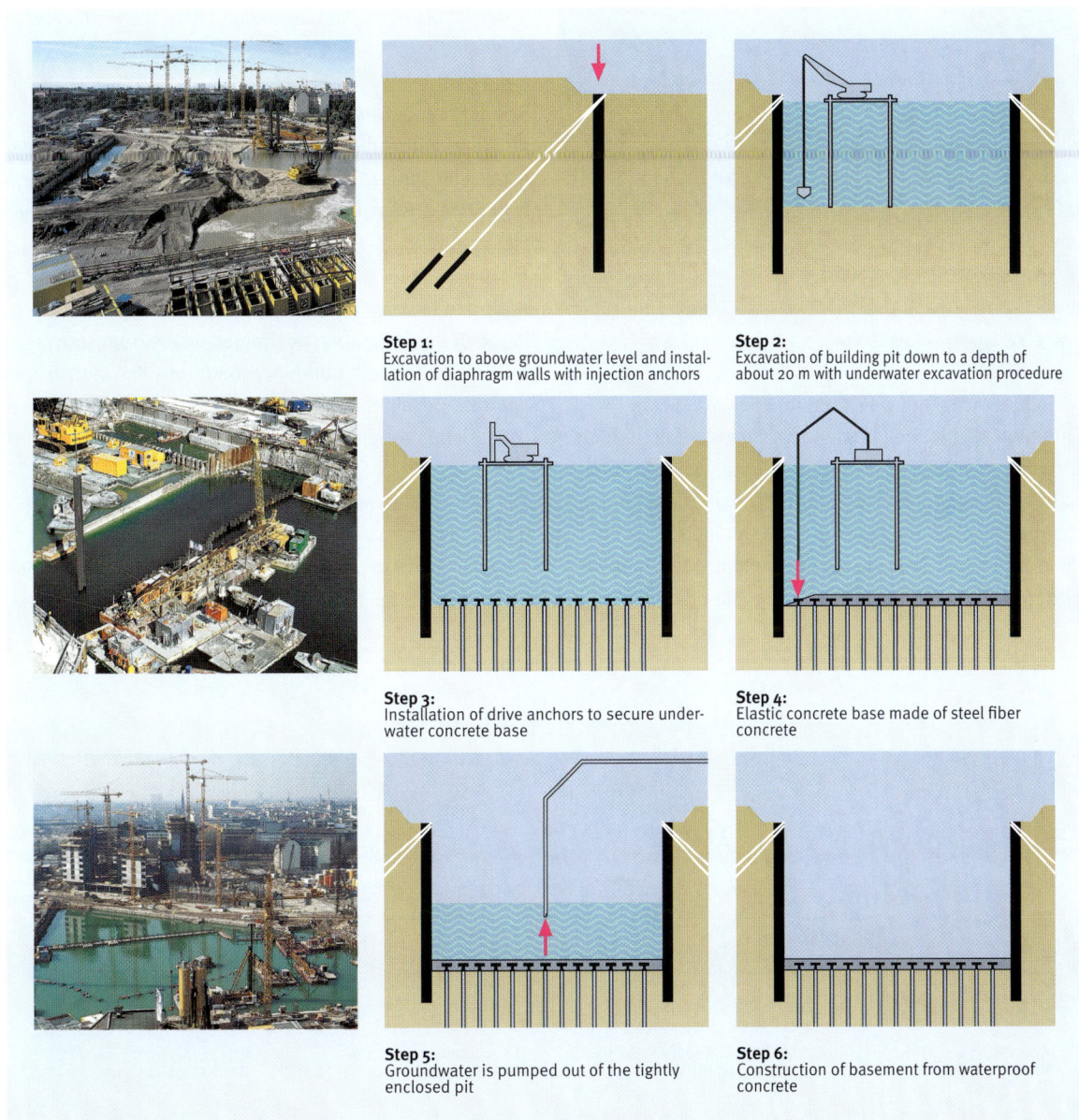

Step 1:
Excavation to above groundwater level and installation of diaphragm walls with injection anchors

Step 2:
Excavation of building pit down to a depth of about 20 m with underwater excavation procedure

Step 3:
Installation of drive anchors to secure underwater concrete base

Step 4:
Elastic concrete base made of steel fiber concrete

Step 5:
Groundwater is pumped out of the tightly enclosed pit

Step 6:
Construction of basement from waterproof concrete

Fig. 3–5 Building pit with diaphragm wall and underwater concrete (underwater excavation)

3.2.2 Foundation Works

Once the building pit base has been completed on a sufficiently large area (ca. 400 m²) foundation work and drainage pipe work can commence. Foundations and drainage pipes are almost always mutually dependent so that their implementation needs to be planned

Fig. 3–6 Production of base plate

simultaneously (but watch out for special cases!). Often, parallel to the foundations, concreted floor ducts, elevator underpasses (hydraulic pillars), wells etc. need to be installed. After completion of these components, the resulting hollow areas are filled (for instance with sieve refuse). Afterwards, a granular subbase is applied and, on this, a reinforced base plate installed.

At least three weeks prior to commencement of the foundation works, the complete, verified planning documents need to be available. This includes especially:

– Foundation formwork plan (with foundation earther)
– Drainage plan
– Foundation reinforcement plan (with connection reinforcement for supports and walls)
– Formwork plan basement
– Reinforcement plan basement (if floor ducts etc. are present)

Special foundation with underpinning: Underpinnings of buildings with subsequent extensions are a special case, especially if this is still required in the groundwater like, for instance, was the case for Weinhaus Huth at Postdamer Platz in Berlin.

High-rise foundations: The concept of a high-rise foundation is primarily determined, aside from subsoil conditions, through boundary conditions like plot size, adjacent developments, and central or de-central rising construction for the foundation.

The easiest and most economical solution is a flat foundation, as long as sufficiently supportive soil layers are available at low depth. Aside from classical pile foundations that transfers structure load to a lower supportive soil layer,

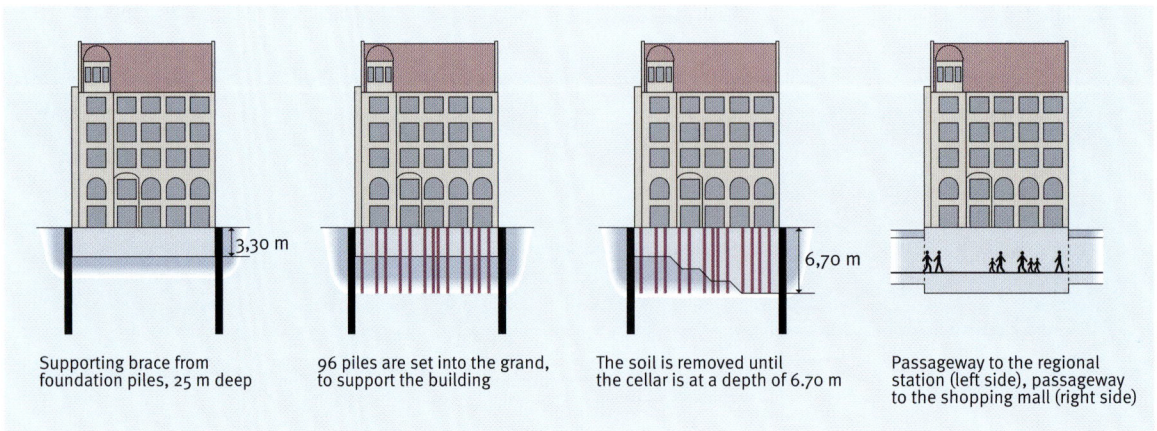

Supporting brace from foundation piles, 25 m deep	96 piles are set into the grand, to support the building	The soil is removed until the cellar is at a depth of 6.70 m	Passageway to the regional station (left side), passageway to the shopping mall (right side)

Fig. 3–7 Excavation underneath an existing building with foundation and fitting out

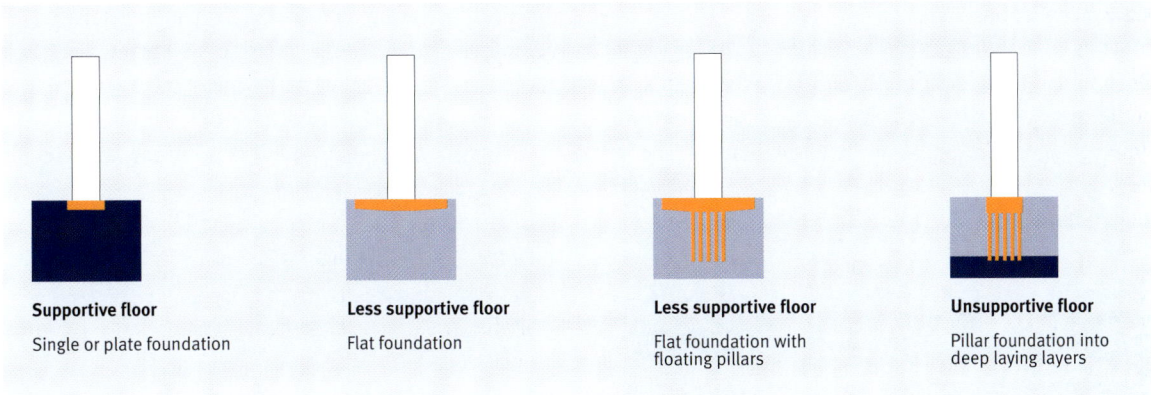

Supportive floor	**Less supportive floor**	**Less supportive floor**	**Unsupportive floor**
Single or plate foundation	Flat foundation	Flat foundation with floating pillars	Pillar foundation into deep laying layers

Fig. 3–8 Foundation process in high-rise construction

"floating pile foundations" are provided if less supportive soil layers are at hand. With these kinds of pile/plate foundations that are used, for instance, for subsoil like Frankfurt clay, building loads are proportionally borne by the foundation plate and the piles. Since the piles transfer their load proportions to deeper soil layers, total settlement can be significantly reduced, which at the same time reduces titling of the building. While for high-rises in Frankfurt, for pure flat foundations, settlements of up to 30 cm have been observed, settlement can be reduced to between 10 and 15 cm if additional piles are arranged for about 30 – 50 % of building load.

For the client, topics like foundations are initially not of interest. But "much money is being buried", if the most economical and technically viable solution is not found in good time. And: cheap can turn out to become expensive, which is the case when settlements that are too large or not uniform have an effect on the supporting structure by means of deformations, which in turn influence extensions and constructive connections. Here, the rule applies: safety comes first!

3.2.3 Rising Construction

During the construction process of the building shell, we distinguish between three different production manners in essence:

– Cast-in-place concrete construction
– Brickwork construction
– Mixed construction (Reinforced concrete – brickwork)
– Prefabricated component construction
– Steel construction

For large building and administration buildings, cast-in-place concrete construction, especially, has become the method of choice, partially in combination with brickwork construction. For large flat construction projects, in as

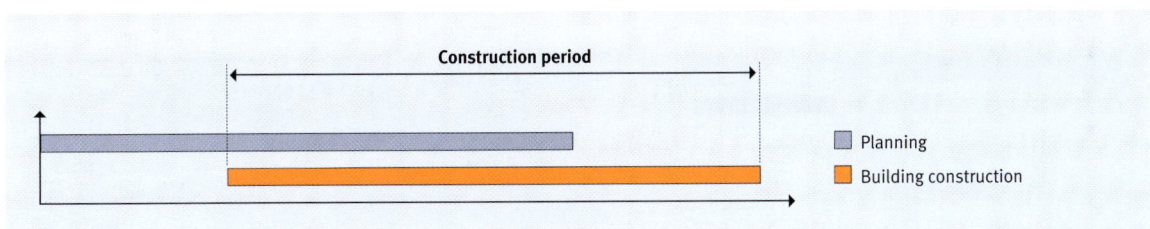

Fig. 3–9 Manufacturing with reinforced concrete, brickwork, mixed construction

Fig. 3–10 Cast-in-concrete construction

Abb. 3–11 Reinforced steel skeleton construction

much as this still happens, mixed construction and brick-work construction are the main methods of choice.

These construction approaches allow for a relatively large overlap between implementation planning and actual construction, since determination of gaps, slits and connecting components can always be undertaken, directly prior to execution of the various individual segments. As a rule, these components are also rather accomodating when it comes to later changes (additional breakthroughs, drills etc.). Completion time, however, is rather long.

In prefabricated component construction and steel construction, we find application areas mainly in industrial construction. The manner of production, especially, influences the process of planning and construction itself.

In order to remain competitive on a price level, reinforced concrete prefabricated components and/or steel construction elements must be dimensioned very precisely from a calculation point of view; also, the number of different types of elements is to be restricted as much as possible.

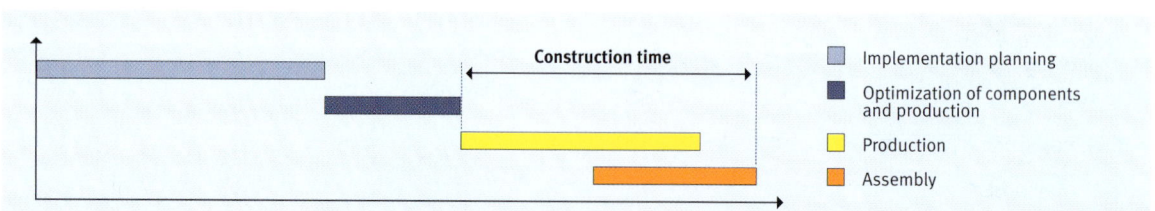

Fig. 3–12 Shell construction, with concrete prefabricated components or steel construction elements

Fig. 3–13 Steel assembly construction

This is only possible when execution planning has been completed in its entirety, allowing for optimization of all the elements. Since, additionally, equal elements need to be manufactured in one production line (change of casings and/or machines), an optimization of production sequence is also required, in coordination also with the assembly process.

Fig. 3–14 Large prefabrication degree during steel assembly construction

This results in a comparatively large planning lead time for this procedure, which also needs to be very well protected since later changes can only be undertaken with enormous effort and, following assembly, are virtually impossible.

Actual assembly time, however, is extremely short, and this can be a decisive factor for the relevant surrounding conditions.

In industrial construction, this process can generally be very easily applied due to the comparatively simple building structures prevailing there.

High-rise construction: One decisive stress factor that can be left out when it comes to construction of lower buildings is rising wind pressure as the height of a building increases. Readings at Frankfurt/Germany, for instance, recorded wind speeds over an extended time period for the Main Tower at a height of 51 m, and for the Commerzbank (Commerce Bank) at a height of 275 m. This showed that a height of 51 m is still heavily influenced by the city and thus presents less wind velocities. A measuring point of 275 m, on the other hand, is comparative to open land.

In principle, one can imagine a high-rise as a kind of fixed rod. This rod is attacked by the wind and, without sufficient reinforcement, would undergo significant bending periods and deformation. This deformation is prevented by reinforcements, which are achieved by various measures.

The initial development period was characterized by story framework in the steel construction manner, which was partially reinforced by non-supporting massive facades. The best known examples for this construction type are the Chrysler Building and the Empire State Building, the latter with the considerable height of 381 m. A disadvantage of this system is the immense constructive effort required at greater heights, which also strongly restricts usefulness of the ground plans. Typical here is the phased, setback manner of construction at greater heights. In Germany, story framework for buildings of up to ca. 40 stories used to be make from sectioned or intercoupled reinforced steel plates.

Elaborate production requirements and restriction of usefulness then led, as a consequence, to Reinforcement with Core. In this type of system, the building core as box girder handles the entire reinforcement, as a rule by means of reinforced steel or composite construction. A big advantage of this construction type is in the great freedom of ground plan and facade design, also as far as outer contours are concerned. However, pure Reinforcement with Core is restricted, owing to comparatively low dimensions, to a height of 120 m to max. 170 m.

Therefore, the so-called Tube systems were developed for higher buildings than that, which shift the box girder to facade level, achieving geometrically much higher stiffness. The box girder is either formed through the spatial frame of a perforated facade or through comparatively narrowly spaced supports with cross-beams. In matters of stiffness, this system behaves in comparison to Reinforcement with Core like a thick pipe versus a thin one. A disadvantage of the Tube system is in the requirement that the building cannot be sectioned at height (Example: Former World Trade Center in New York or Amoco in Chicago). Further, accessibility of entrance levels is a constructive problem since the narrow spacing of the supports must be absorbed from the upper stories via complex interception structures and directed to the foundation.

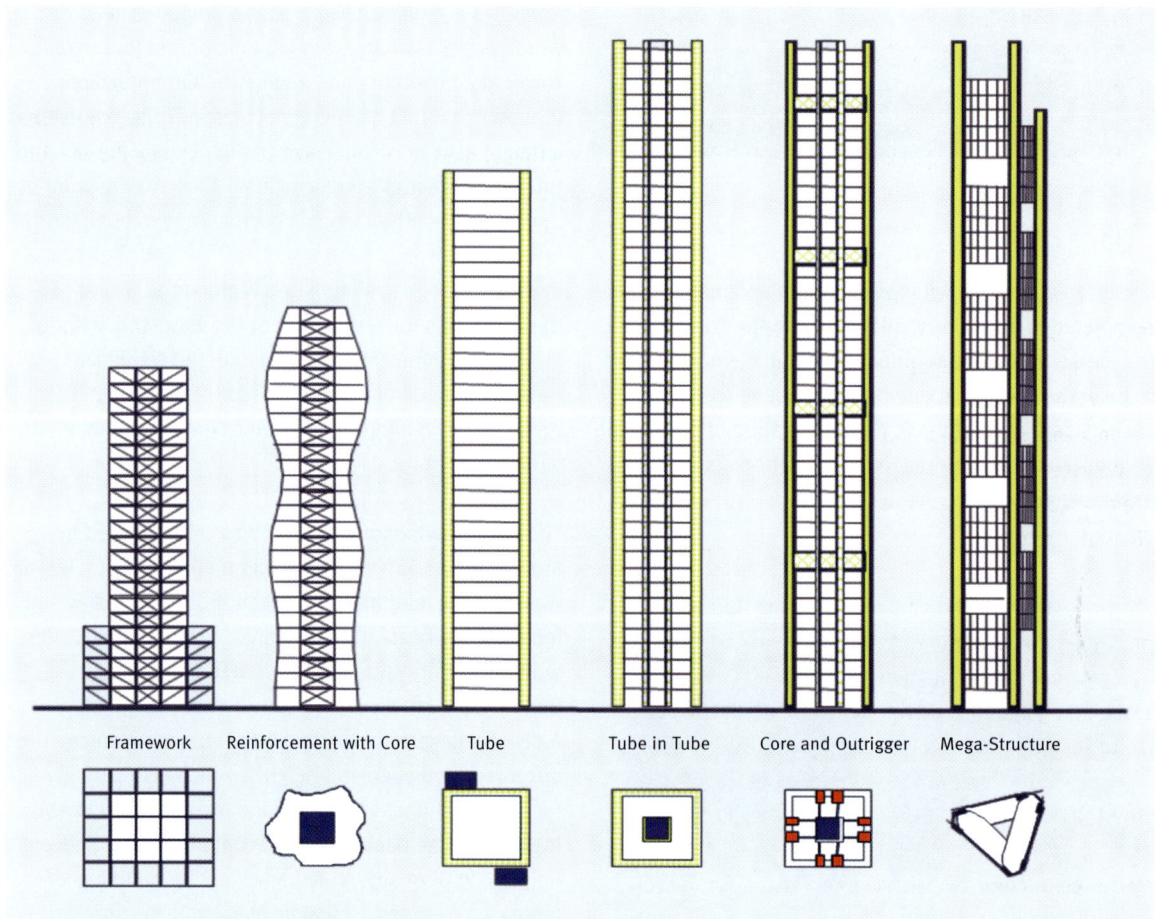

Fig. 3–15 Construction principles for high-rise buildings

A further development here are the so-called Tube-in-Tube systems, where both core and facade or designed as hollow girders, something which greatly reduces shear deformation of the system. Load distribution between facade and core is about half each, something that has a positive effect on economical dimensioning of the components. In the Tube-in-Tube, the support system of the inner and outer tube is completely decoupled, something that provides greatest possible freedom for the use of the interior space (Example: Measuring tower Frankfurt and World Financial Center in Shanghai at 460 m).

A further development for buildings of great heights is the so-called Core and Outrigger system. This support structure consists of a few mega-supports in the facade region and of the core, by means of which the mega-supports on the engineering levels are connected by shear-strength. Through restricting the connection to the engineering/technical floors, use in the office floors continues to be possible without restriction while, at the same time, there is increased stiffness for the entire system (Example: Jin Mao Building, Shanghai).

Finally, there is also the so-called Mega-Structure principle where a trans-story Framework system is combined with mega-supports. The first acknowledged example for this support structure form was the Bank of China at Hongkong, where coupling of the mega-supports is undertaken via a giant framework in the facade as well as in the layout diagonals. A current example would be the Commerce Bank at Frankfurt, where the mega-supports are connected by shear-strength through multi-floor vierendeel girders.

Maximum possible heights for these construction principles do not seem to have been reached to this day. Designs ready for implementation for buildings up to 1.000 m high, are available – but whether this would make sense is doubtful.

Construction method: Generally, when it comes to construction methods, we distinguish between concrete and steel construction. For the very high buildings in the United States, usually two stories a week are assembled in steel. In this, all components are already prefabricated in the shop, welded as segments. On site, as much as possible, only screw connections are used. For the reinforced steel buildings in Germany, as a rule, it is only possible to handle one story per week, whereas construction speed must always be considered in connection with possible extensions. Saving construction time seems to be a more important issue in the USA than, for instance, in Germany and this, of course, is also very heavily dependent on the extent to which use of the building immediately after completion is even warranted. Concerning ceilings, composite construction has become the method of choice, where implementation of ceiling boarding is not required and, therefore, another transport route can be saved.

Overall, it can be stated that the construction methods currently used in Germany and throughout Europe have an average production time of one story per week.

For trade fair tower in Frankfurt, however, there was a special, coordinated procedure between slip form of the cores and climbing formwork of the outer walls, which led to a construction speed of ca. 1.5 stories per week, in general.

3.3 Building Envelope

3.3.1 Carpentry Works

Increasingly, sloping roofs are used for large buildings, so that carpentry works have become ever more significant. In essence, services extend to:

– Creation of substructure
– Creation of roof boarding and articles

For certain types of objects like gymnasiums, swimming pools and large assembly halls, entire structures are also executed as room sides like:

– Wall constructions
– Ceiling and roof supports and trusses

These structures are usually executed as glued supports, beams and trusses, often also in connection with steel cable bracings.

Notes for process planning:

– Prior to commencement of carpentry works, all connection and bearing points of the constructive shell work must display the required calculated strength
– Roof sealing should come immediately after the carpentry works in order to avoid excess soaking of the wood structure

3.3.2 Roof Sealing and Plumbing Works

Due to their close connection, these two trades need to be executed jointly; it is best if they are part of a call for tender in the form of a BOQ. We distinguish between flat roofs and sloping roofs.

Flat roofs: As a rule, flat roofs are nowadays only designed as warm roofs. For one, this is more economical and, secondly, materials have become much more resistant nowadays, if they are selected carefully.

For a regular design **warm roof**, the roof skin adjoins directly to the concrete ceiling. In order to prevent soaking of the insulation from below, a vapor blockage is therefore required.

The foam glass design version is somewhat more expensive but also more durable and less effort to maintain. This version should always be chosen when, instead of gravel filling, a walk-on plate cover or roof vegetation is intended.

If the roof is not to be walked on, the principle of the so-called "inverted roof" applies, where the sealing level lies underneath the insulation itself. This is then to be designed from extruded polystyrene with system approval.

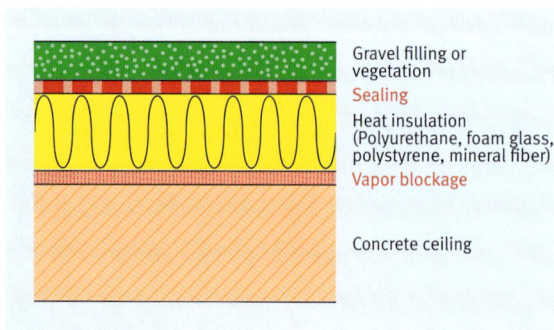

Gravel filling or vegetation
Sealing
Heat insulation (Polyurethane, foam glass, polystyrene, mineral fiber)
Vapor blockage

Concrete ceiling

Fig. 3–16 Flat roof as warm roof

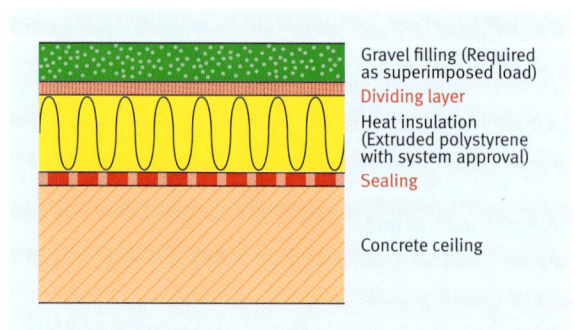

Gravel filling (Required as superimposed load)
Dividing layer
Heat insulation (Extruded polystyrene with system approval)
Sealing

Concrete ceiling

Fig. 3–17 Flat roof as a cool roof (Inverted roof)

Sloping roofs: In contract to a flat roof, sloping roofs are almost always designed as cold roofs with a interim space for air circulation (rear ventilation).

At the example of a tile roof, we can see how, under the actual roof skin (tiles), there is an air vacuum. Although, in principle, no vapor blockage is required for cold roofs on account of rear ventilation, as a rule it is nonetheless installed for security reasons. Further, vapor blockage is required for a cold roof located above air-conditioned rooms due to the high level of vapor pressure.

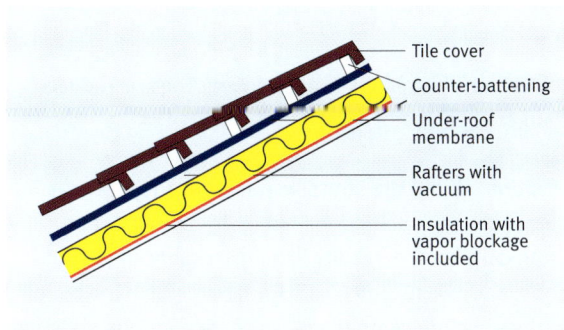

Tile cover
Counter-battening
Under-roof membrane
Rafters with vacuum
Insulation with vapor blockage included

Fig. 3–18 Tile roof as cold roof

Important for flat roofs is also the shape of the edge rims as they go over into the facade. For this, any of the options shown in the image can be chosen.

3.3.3 Traditional Facades

The facade seals off the vertical parts of the building envelope with enclosed (opaque) and lit (transparent) surfaces. For residential construction, masonry and concreted perforated facades with supportive function are common, in connection with windows of wood, plastic or aluminum. When it comes to office building construction, the majority of facades are non-supporting and pre-assembled, made from wood or metal, where the enclosed and lit surfaces form a constructive unit.

Function: Generally, facades can be distinguished according to their function, into:

– Supportive facades
– Non-supportive facades

Supportive facades are generally made from brickwork or concrete steel. As a system, together with the inbuilt windows, they simultaneously serve for load absorbing of the ceilings and also as climate protection. They are constantly being created as shell work progresses and are later plastered or paneled.

Position: Non-supportive facades have only one function, which is climate protection, and therefore they can be installed independently of shell construction within a certain follow-up period. We distinguish between non-supporting facades according to their position.

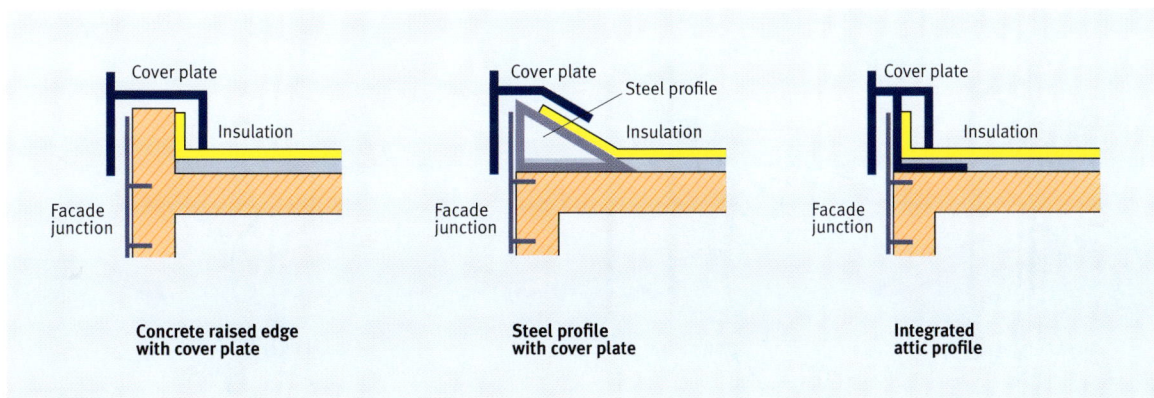

Cover plate
Insulation
Facade junction
Concrete raised edge with cover plate

Cover plate
Steel profile
Insulation
Facade junction
Steel profile with cover plate

Cover plate
Insulation
Facade junction
Integrated attic profile

Fig. 3–19 Alternative attic shapes

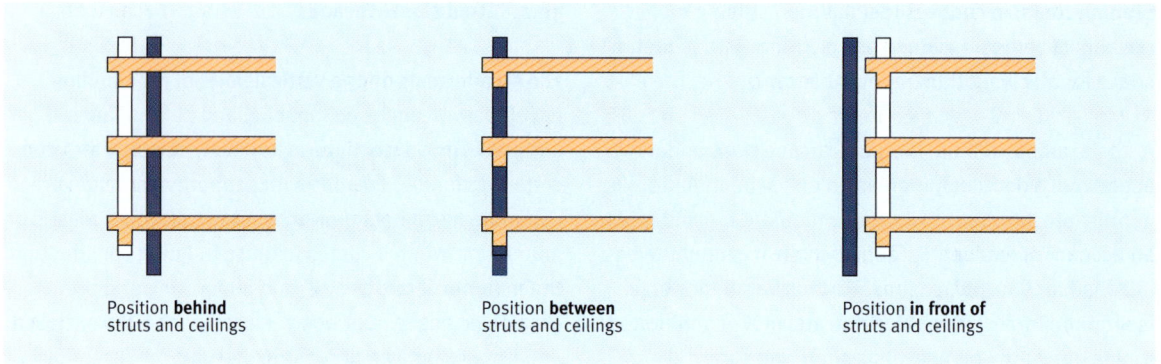

Position **behind** struts and ceilings

Position **between** struts and ceilings

Position **in front of** struts and ceilings

Fig. 3–20 Alternative facade levels

Position in front of struts and ceilings: This facade is also known as a curtain wall, it encloses the entire building in its vertical spatial enclosures and handles all building physics functions.

As a rule, it is designed as a metal facade with integrated windows and solar protection devices. Assembly is undertaken following completion of shell construction, since there is great danger of damage.

Position between struts and ceilings: This variety is most often used when a reinforced concrete skeleton structure is "filled in" with brickwork. However, simple design adjusted facade elements of wood or metal are also installed in this position. In these cases, the frontal sides of the ceilings and struts need to be separately insulated. Facade installation can be under-taken shortly after shell construction. There is reduced danger of damage.

Position behind the struts: In this case, the facade is clearly set back behind the struts. This construction approach, for one, results in unnecessarily large strut gaps, and, for seconds, cold bridges result in the region of the puncture component, which then require elaborate "wrapping" of construction components lying outside the facade. Assembly can be undertaken immediately following stripping of a ceiling. The setback position constitutes virtually no danger of damage.

Structure: Clear distinctions can be made in terms of structure, also:

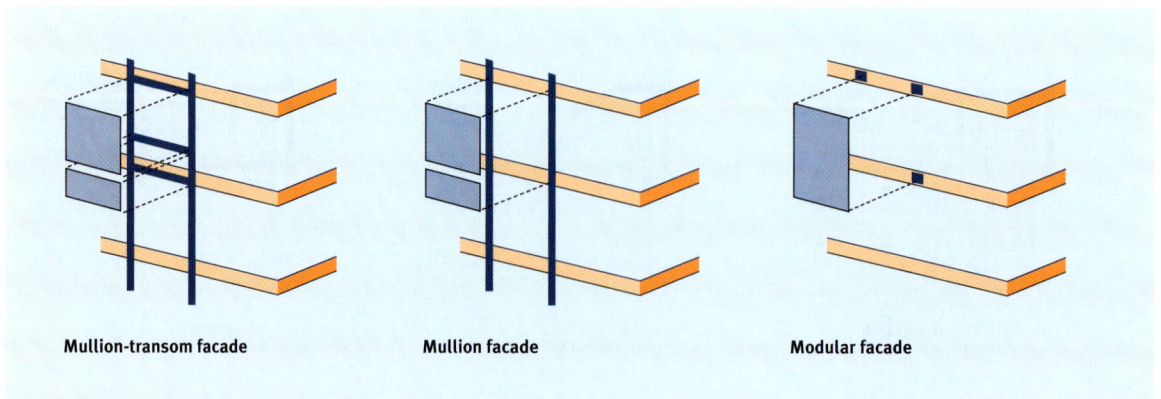

Mullion-transom facade

Mullion facade

Modular facade

Fig. 3–21 Alternative manners of facade construction

Mullion-transom-facade: The facade substructure consists of vertical mullions and horizontal transoms. This means that small-part components can be used. Owing to the low degree of prefabrication, even small metal construction firms can offer this service. Assembly speed is quite slow.

Mullion construction: Horizontal components are not needed for the substructure, which means that the elements need to be larger. Prefabrication degree rises, and assembly speed is faster by 20 % to 30 %. Small metal construction firms can just about still handle this construction manner, and in case of being assigned the job, they often obtain the elements elsewhere and merely assemble them.

Element structure: Facade elements are fully pre-fabricated components, which as a whole are directly fastened to the support structure via anchor plates. Required stiffness can only be obtained by means of special structures, for instance

– Integrated frames
– Forming and/or profiling of metal sheet plates
– Formation of panes

Only very few specialized manufacturers with their own engineering departments are capable of producing such elements in good quality. Over the course of shell construction, the anchor plates are built in, too. Assembly speed for the elements is about five to six times higher than for a mullion-transom structure.

Installation process: Facade builders prefer assembly from top to bottom or vice versa, since the elements are easier to align during this process.

Frequently for high buildings, however, facade construction needs to come after shell construction, meaning that horizontal assembly cannot be avoided.

Notes for process planning:
– Aligned facades and brickwork can follow immediately after shell construction, provided that the respective protective measures are in place (Spacing three to four floors).
– Assembly of metal curtain walls should only commence once shell construction works above are completed (except for substructure).
– For facade assembly of light facades, scaffolding must be provided (Watch out in case of roofs lying below, these must not be finally insulated before scaffolding is taken down again, otherwise there is danger of damage).
– Light facades with so-called maintenance balconies can be mounted without scaffolding.
– In the case of supporting brickwork facades with cladding (Plaster, slates, metal clinkers), initially, the windows including glazing can be installed, facade cladding can then follow (time advantage).
– Welding works at the rim area (e.g. heating) must be completed prior to commencement of glazing works, due to danger of damage.
– If the schedule does not allow for glazing to be undertaken in due time, the shell construction BOQ must make allowance for provisional foil glazing.

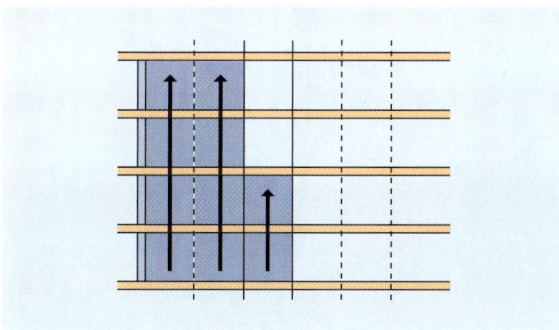

Fig. 3–22 Facade assembly vertical

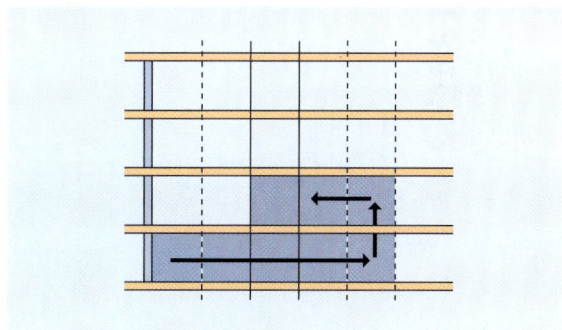

Fig. 3–23 Facade assembly horizontal

3.3.4 High-Rise Facades

One decisive factor for overall construction time is the time frame in which facade assembly follows shell construction. Fast timeframes are essentially achieved here by pre-assembling the facades as complete elements, attached from the interior of the building by means of special crane trolley facilities without outside scaffolding. For this, the facade elements are completely equipped with windows and outer cladding as well as the required dividing wall and ceiling junctions. Even natural stone cladding and radiator cladding is integrated into the elements prior to assembly.

3.3.5 High-Tech and Energy Facades

In future, the facade of an environmentally friendly, climate-adjusted house will become variably adjustable to climate – independent of season or time of day.

Owing to improved usability and intelligent control of energy flows of light and heat, it will be able to significantly reduce energy consumption and at the same time noticeably improve comfort levels inside the rooms of our buildings.

Primarily, variable facades include molecular facade systems with electrochromic and thermochromic glass, photovoltaic cells, holographic layers, but also double-skin facades. While everything points to economical concepts not being able to be realized in the near future for molecular facade systems, there is, even today, already a developing of economical solutions for a double-skin facade, provided that certain framework conditions apply.

In a double-skin facade, a second skin made of safety glass, capable of being rear-ventilated, backs a conventional facade. Hence, there is an outside and inside facade. Generally, there is an adjustable solar protection device situated between inner and outer facade. The windows of the inner facade are generally capable of being opened, meaning that natural ventilation of the rooms above the facade interim space is possible. In this, outside air is supposed to flow into the room through the lower half of the window, waste air should stream out through the upper half.

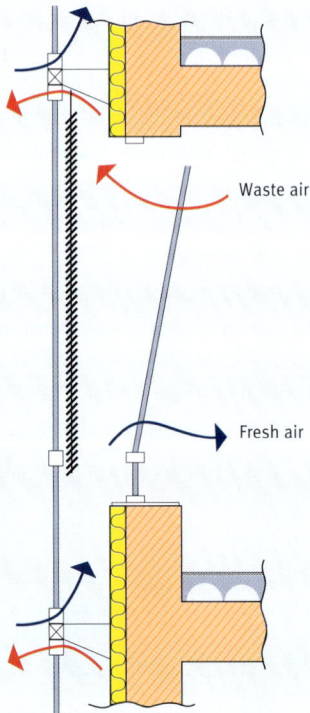

Fig. 3–24 Principle behind a double-skin facade with natural ventilation

Fig. 3–25 High-rise facades with natural ventilation

3.4 Building Services Equipment

The individual trades of technical equipment are looked at in the chapters to follow. These are the groupings that can be undertaken:

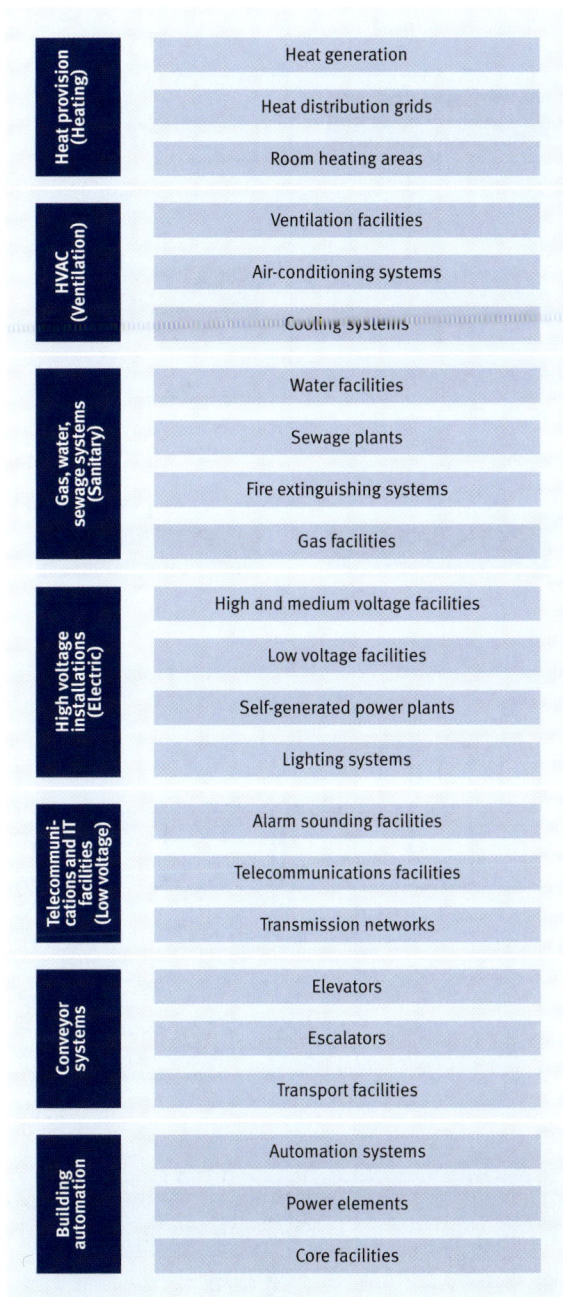

Heat provision (Heating)	Heat generation
	Heat distribution grids
	Room heating areas
HVAC (Ventilation)	Ventilation facilities
	Air-conditioning systems
	Cooling systems
Gas, water, sewage systems (Sanitary)	Water facilities
	Sewage plants
	Fire extinguishing systems
	Gas facilities
High voltage installations (Electric)	High and medium voltage facilities
	Low voltage facilities
	Self-generated power plants
	Lighting systems
Telecommunications and IT facilities (Low voltage)	Alarm sounding facilities
	Telecommunications facilities
	Transmission networks
Conveyor systems	Elevators
	Escalators
	Transport facilities
Building automation	Automation systems
	Power elements
	Core facilities

Fig. 3–26 Trades of technical equipment

The first three groups are commonly known as "HVS" or "Heating, Ventilation, Sanitary" trades and generally are handled by one single contractor. There are totally different requirements for "Electric power and low voltage facilities", which constitute the second group of trades. Separate, once more, are the executing firms for conveyor systems and building automation. Since for planning, as well as for assembly and control, there are numerous interfaces between the various trades, a bundling up in form of working units is often desired – as a rule, however, with exception of conveyor systems.

A standard process, which can always be applied, does not exist. In individual cases, rather, it depends on the components of technical equipment, and their inter-meshing with extensions, how the works are specifically undertaken. Some ground rules, however, will apply with great probability for most procedures and they are illustrated at the example of an administration building in Fig. 3–27.

Before installation works can commence, reinforced concrete works above the story to be installed need to be concluded and provisional – but safe – sealing needs to be in place. These two requirements are an absolute prerequisite before the start of rough assembly for ventilation and elevators.

Prior to commencing electric cabling, insulation works and substructures for suspended ceilings, any possibility of water entry into the working area must be excluded. For this, roof insulation should categorically be in place, the facade ought to be closed in the relevant areas and all water-conducting pipes should be compression tested. Otherwise, very cost-intensive water damage cannot be excluded, which can also result in considerable schedule delays.

Following insulation and laying of the main electric arteries, wall slits and ceiling breakthroughs can finally be closed. Only then, final fitout works may commence, like floor installation, floor covering works and closing of the suspended ceiling.

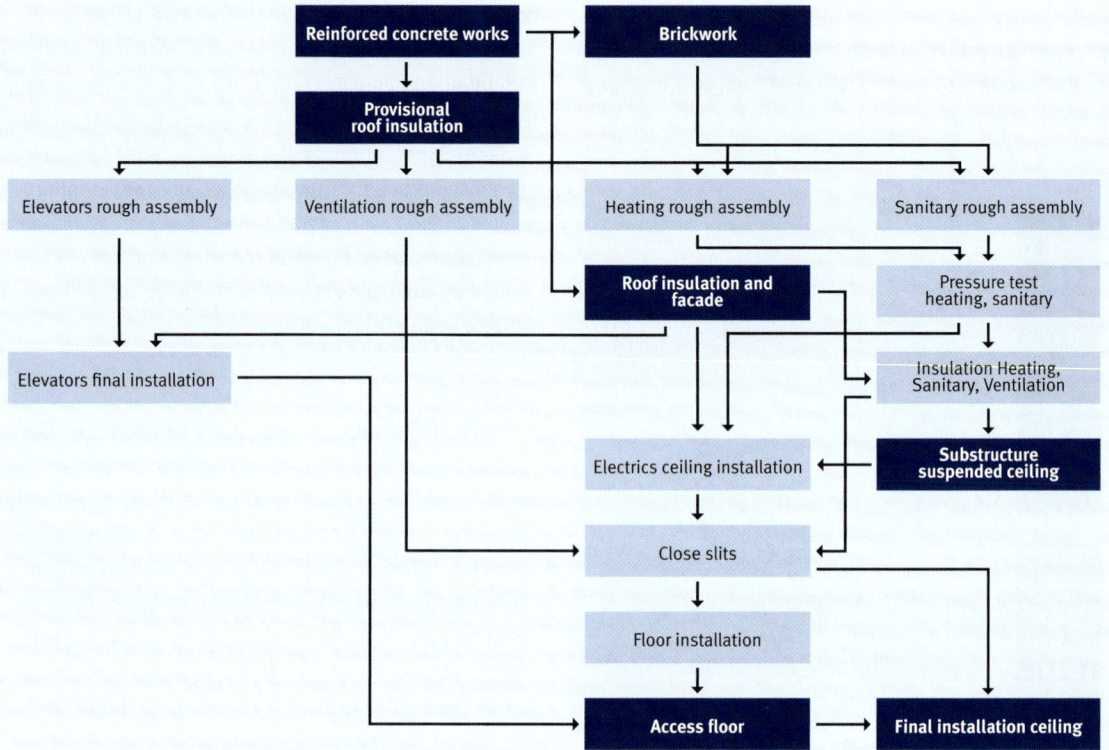

Fig. 3–27 Scheme process technical equipment

For multi-story buildings – especially for high-rises – one will not be able to wait until reinforced concrete works have been concluded but, for instance, after four or five floors, ceiling openings are going to be provisionally sealed and then rough assembly will commence (see Fig. 3–28). Due to generally required, temporary ceiling bracing, reinforced concrete works need to be undertaken a minimum of two to three floors above the sealing level.

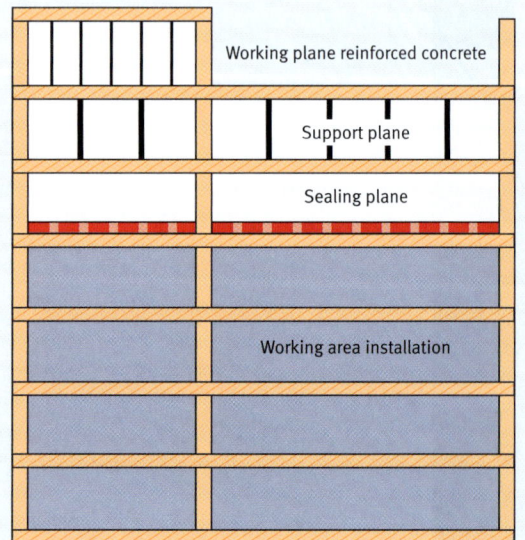

Fig. 3–28 Fitout segments for tall buildings

3.4.1   Sanitary Engineering

Sanitary engineering, in essence, deals with gas and water supply as well as wastewater disposal of buildings

and terrains. For better understanding concerning distinction of public and/or community gas and water supply and disposal facilities, the following illustration may be of assistance:

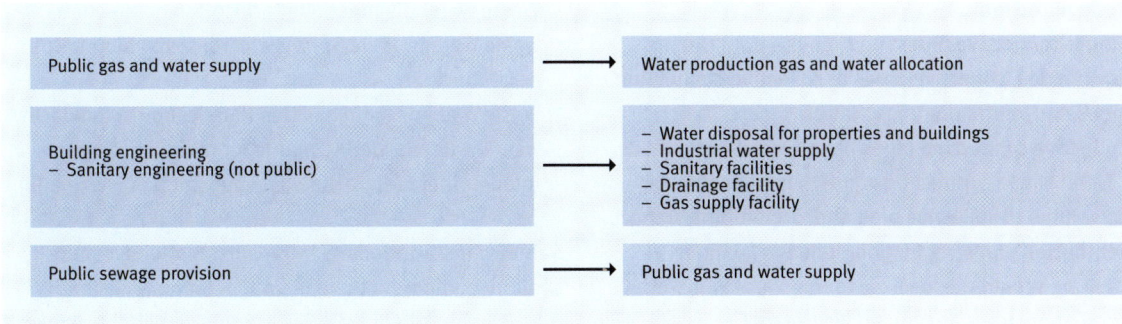

Fig. 3–29 Allocation diagram water and waste water

Water and wastewater inside the building: Drinking and industrial water is guided from the house connection to the different consumption sites via the risers. Each consumption site also includes a wastewater drainage point. For wastewater, as is also shown in the illustration to follow, we distinguish between the following line segments:

– **Connecting duct** guides rainwater, wastewater or mixed water from the property limit and/or the first purification duct to the public sewage canal.
– **Main conduits** are either located inside a building or in the earth underneath the foundations. They guide rain or wastewater to the connecting duct.

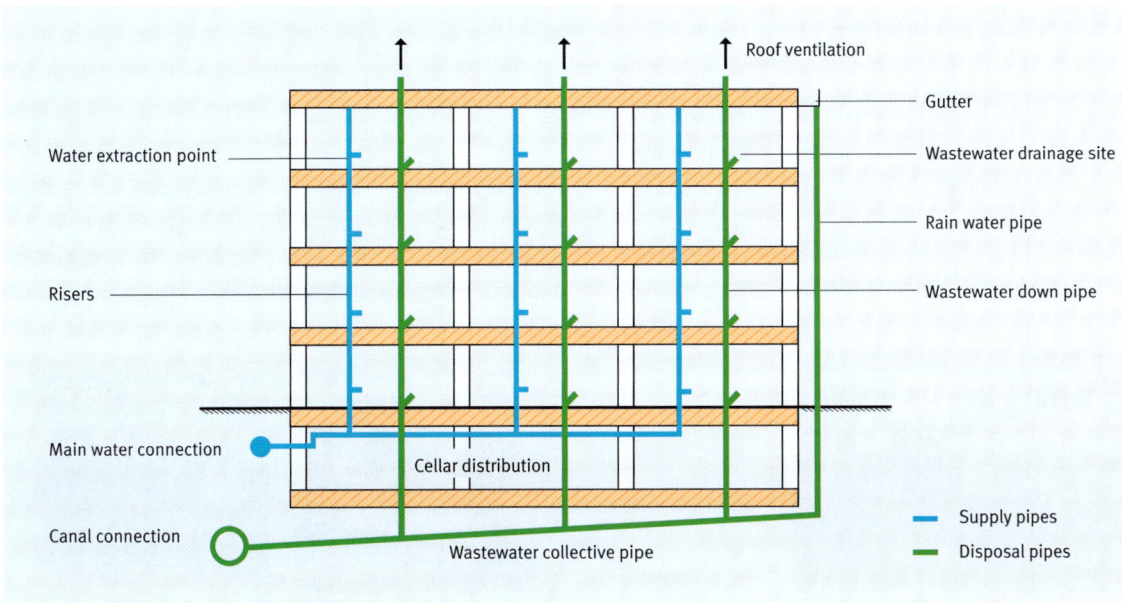

Fig. 3–30 System behind sanitary installation

- **Collective pipes** gather water from down pipes and connecting pipes.
- **Down pipes** for wastewater are interior, vertical and sometimes also warped pipes, which run through several floors and, at the upper end, are ventilated through the roof.
- **Single connective pipes** run from the odor trap to down pipes, collective pipes or collective connection pipes and/or the pumping station.
- **Collective connection pipes** absorb wastewater from several single connections and then guide it to the down pipe or collective pipe and/or pumping station.
- **Ventilation pipes** ventilation drainage facilities. They are required especially for pressure adjustment. Ventilation pipes are guided via the roof.
- **Rainfall pipes**, whether interior or exterior version, divert the rainwater.

For the purpose of roof draining, separate down pipes are required. Drainage of wastewater is different from community to community. We distinguish here between:

- Mixed system: All wastewater is drained into a joint canal.
- Separate system: For home-generated wastewater and rain water (roof drainage), separate ducts are provided.

Specialist waste water systems: Special installations become necessary when wastewater is exceptionally polluted, for instance:

- Grease separators (for grease-charged waste water from butcher shops, industrial kitchens etc.)

- Fuel separators (Mechanical workshops, garages with car wash provisions)
- Heating oil separators (Oil heating control rooms)
- Starch separators (Industrial kitchens with potato peeling machines)
- Waste water disinfection (Purification of infectious waste water from hospitals etc. through chlorination or heating)
- Waste water decontamination (Purification of radioactive waste water in cooling plants for radioactive waste – through chemical treatment, evaporation or incineration)

Sanitary installations: Sanitary installations include sanitary equipment to be installed, for instance bath tubs, washbasins, lavatories, urinals and kitchen sinks with all corresponding valves and fittings.

Fittings like mirrors, bathroom cabinets and bathroom furniture, shelves, towel holders, soap holder and paper holders are included in sanitary installations.

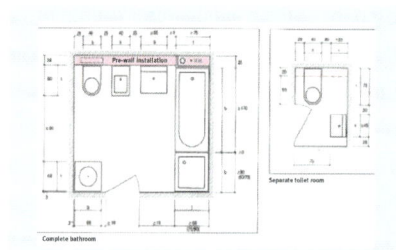

Fig. 3–32 Layout with equipment

Process variants: To simplify assembly, prefabricated assembly blocks are increasingly being used. These are produced in the workshop and only need to be mounted and connected anymore on location.

Fig. 3–31 Grease separator

Abb. 3–33 Assembly blocks

3.4.2 Fire Extinguishing Systems and Protective Devices

Inside buildings, water hydrants are installed that serve for fire fighting in case of fire breaking out. As a rule, these can be divided into:

- Dry firefighting water facilities
- Wet firefighting water facilities
- Wet/dry firefighting water pipes
- Manual fire extinguisher systems (e.g. Fire extinguishers)
- CO_2 fire extinguisher systems (Fire extinguishing gas)

Firefighting water pipes and wall hydrants are to be placed in the fire section (staircase) and to be insulated when guided through outside fire sections F 90. Whenever wall hydrants are fitted into wall niches, this must not power the fire resistance category of the wall.

Dry firefighting water pipes: DIN 14 462 uses the expression "Risers, dry" whenever firefighting water, in the event, is to be supplied by the fire station. This allows for access to fire fighting water without time delay that results from laying of hosepipes.

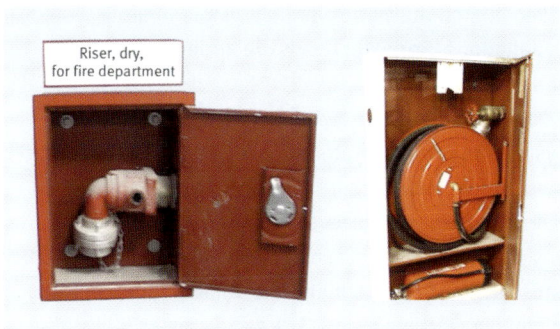

Fig. 3–34 Point of withdrawal for firefighting water and wall hydrant

Water feed can be undertaken following turn-on of a firefighting pump of the fire station via hydrants from public drinking water supply or it can be undertaken with non-drinking water from natural water sources.

Wet firefighting water pipes or "Risers, wet" are subjected to permanent water pressure and, therefore, are always ready for operation owing to connection to a public water supply network. Prerequisite here is the arrangement of risers in frost-proof staircase areas.

Ventilation equipment and pipe routes needs to be arranged in a manner to ensure that the pipes are impeccably ventilated in case of commissioning and decommissioning, in the latter case ensuring that the pipes are totally emptied of water.

Sprinkler systems are automatically acting systems with closed nozzles. A sprinkler system is designed for recognizing a fire in its creation stage already and to extinguish it or get it under control to a point that allows extinguishing by other means. Aside from a few exemptions, it ought to cover the entire building.

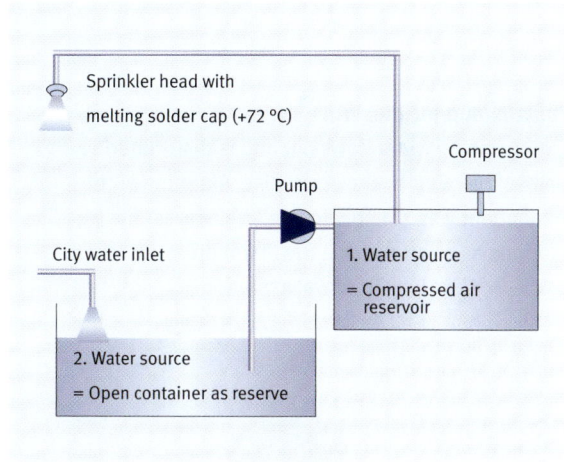

Fig. 3–35 Principle behind the sprinkler system

There are three varieties available for installation of the sprinkler tank:

- Installation prior to shell construction
- Installation after shell construction through assembly openings (Problematic since large dimensions)
- Cellar-welded sprinkler tank (expensive)

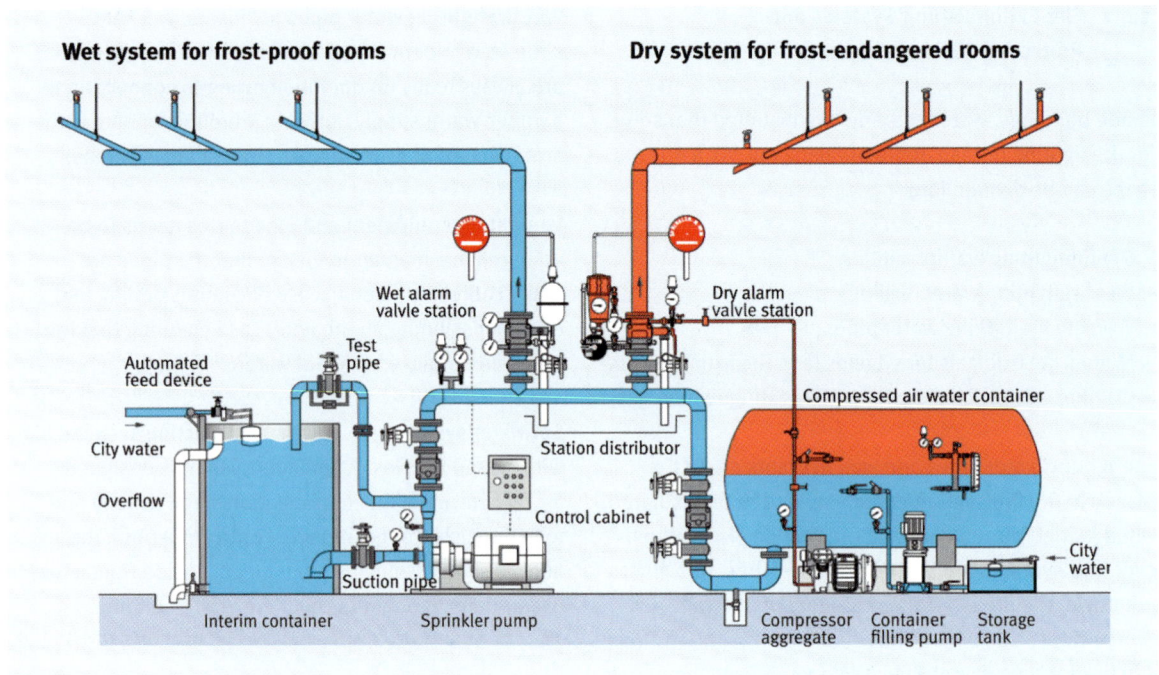

Wet system for frost-proof rooms

Dry system for frost-endangered rooms

Wet alarm valve station

Dry alarm valve station

Test pipe

Automated feed device

Compressed air water container

City water

Station distributor

Overflow

Control cabinet

City water

Suction pipe

Interim container | Sprinkler pump | Compressor aggregate | Container filling pump | Storage tank

Fig. 3–36 Sprinkler system, complete (Source: Minimax)

Functional principle: When the sprinkler is set off, extinguishing water flows out and is distributed across the room by the spray disc. Alarming of the permanently occupied site, e.g. the nearest fire station, is undertaken via the alarm pressure switch of the existing wet alarm valve station.

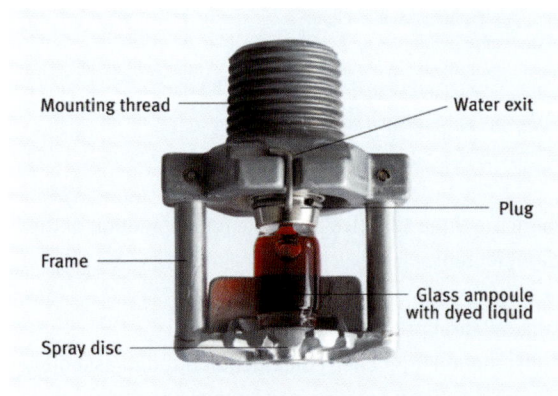

Mounting thread

Water exit

Plug

Frame

Glass ampoule with dyed liquid

Spray disc

Fig. 3–37 Sprinkler head

Installation of the sprinkler pipes, as a rule, is undertaken immediately after ventilation and heating system installations. Sprinkler heads are added last.

CO_2 fire extinguisher systems: The wet sprinkler systems running on water are used especially in department stores, malls, sport halls, industrial facilities and open plan offices. However, they are unsuitable for any setting where, in case of application, they could cause irremediable damage like, for instance, computer centers in machine rooms. The same applies for locations where highly inflammable material could lead to rapid spread of fire, for instance in laboratories or chemical storage places: There, water is not to be used as extinguishing medium. In these cases, water is replaced by inert gas (protective gas), for instance Argon air's or CO_2. The fire is killed through pushing out the air oxygen, without doing any damage.

3.4.3 Heat Supply Systems

Heat supply in structural engineering has two tasks to fulfill:

– Heat transport through static heaters, which transmit heat directly to the room, e.g. radiators, floor heating, ceiling or wall heating.
– Heat transport through dynamic heating, where the heat is made available to the room indirectly, by means of a ventilation duct system and/or a ventilation device.

Owing to legal requirements (Energy savings law, CO_2 reduction etc.), modern devices have been proven to supply very good heat insulation. This is why, for heat supply, pump-driven warm water heating has become primarily the method of choice.

The most energy efficient distribution system is the dual-pipe system, as low-temperature heating, with a maximum flow temperature of 75 °C. Often, flow temperatures are planned for at an even lower level since the lower heating load of the buildings allows for this.

Static heating: Application of radiators is most frequent in the window areas, with the radiators containing circulating warm water. Heat emission takes place in form of radiation and convection.

With the dual-pipe system, as is common nowadays, each radiator is set off parallel to the other, meaning that each radiator has a separate flow and return pipe. This allows for the required temperature ranges to be adjusted to each heating radiator system. The same facility, hence, can handle adjustment of radiators, floor heating and wall/ceiling heating.

Surface heating systems like, for instance, floor heating, ceiling heating (thermal concrete core activation) and wall heating are getting established more and more. They are characterized by lower heating water temperatures, often result in favorable indoor temperature distribution and thus generate a good level of comfort. Another advantage lies in the fact that surface systems (most frequently ceiling and floor) can also be used in summer as cooling surfaces.

In case of glazed, tall halls, facade elements can also be used for heating.

For contemporary low energy buildings that place a high demand on the principle of energy savings, conventional heat generation systems (heat boiler) are used in combination with regenerative systems (e.g. heat pump, solar exploitation).

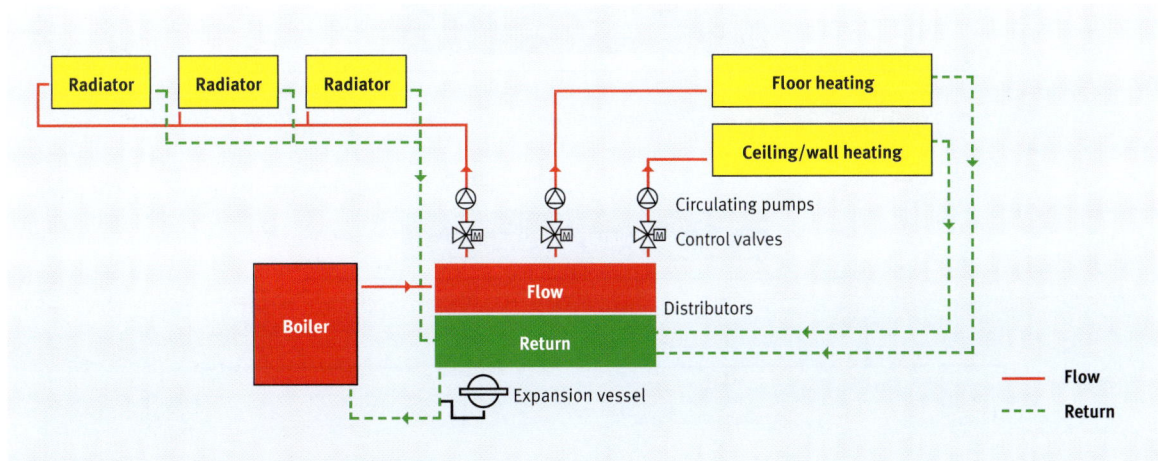

Fig. 3–38 Principle behind a dual-pipe heating system

3.4.4 Indoor Air Technology (Ventilation and Air-Conditioning Technology)

Principally, a building can be **naturally** ventilated via wind pressure or thermics, or **mechanically** via ventilators. We distinguish between natural or mechanical ventilation.

Natural ventilation: Most buildings, up to a room depth of 7,50 m, can be ventilated via windows or air ducts in a natural manner.

Natural ventilation, however, no longer suffices when load in the rooms, caused by people or equipment or room depth, becomes too large. Outside influences also, like noise or pollutant emission levels, can lead to opening of the windows no longer becoming possible or bearable. In these cases, mechanical ventilation is required.

Mechanical ventilation systems: These are distinguished by whether they work exclusively with air or with a combination of air and water. In the later case, water is used for transport of heating or cooling energy to the place of consumption.

There are several air treatment levels for mechanical ventilation. With the assistance of ventilation or air conditioning systems, rooms are not only capable of being ventilated but they can also be conditioned at various levels, cumulating in an air conditioning system. Specifically, we distinguish between the following thermodynamic air treatment methods:

– Filter (F)
– Heating (H)
– Cooling (C)
– Moisturizing (M)
– Dehumidification (D)

A ventilation system that encompasses all these treatment levels and further warrants constant room readings is also known as a comfort air conditioning system.

Air-Only systems

Constant Volume Flow systems (CVF systems): Volume flow remains constant, flow temperature is adjusted according to requirements.

As the name already indicates, volume flow for this system is at a constant level. Reaction to different indoor requirements is undertaken solely on the basis of air conditioning, meaning that the air is centrally heated

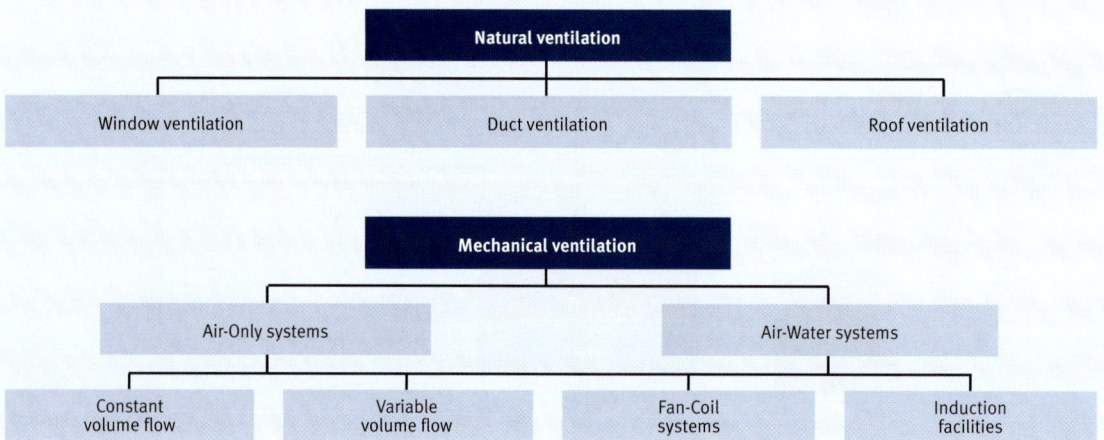

Fig. 3–39 Ventilation systems

more or less, or cooled, humidified or dehumidified. One or more rooms of the same condition result. A requirement here is that air volume flow is low or heat load for the rooms are more or less the same since, otherwise, different room conditions may result.

Usual application: Rooms where air volume flow is determined by the number of people or by the harmful substances discharge requirement, or rooms with constant load, for instance in assembly areas, storage areas, control rooms, in office areas with low cooling loads. Maximum cooling load ≤ 25 W/m² ventilated area for offices.

Advantages:
   Low control effort required for one-zone system
 − Low space requirement for technical installation
 − Low maintenance costs

Disadvantages:
 − Higher energy costs in comparison to VVF system
 − In the case of one-zone systems, no individual
   regulation of room temperature is possible

Variable-Volume-Flow systems (VVF systems):

Principle: Inlet temperature remains constant, volume flow is adjusted according to requirements. For the described facility systems, volume flow is constant and inlet temperature variable, meaning according to requirements within a regulation section it can be adjusted to be higher or lower. The disadvantage mentioned previously, meaning that it is only possible to react to changing heat loads on a zone-by-zone basis and also that the 100 % volume flow requires a greater level of transport energy, does not exist for the VVF system.

Inlet air temperature for VVF systems is kept at a constant level; volume flow is adjusted to loads either zone-by-zone or room-by-room. VVF systems require additional components that allow for a variability of air volume flow.

Change of air volume flow in the devices is undertaken by frequency-dependent regulation of the ventilator drive. For the various zones or rooms, variable volume flow regulators are required on the inlet and outlet air side that use room temperature readings and control units to adjust inlet and outlet air volume. Inlet air, for instance, is kept at a constant level of 18 °C and transmitted to the rooms. In the event of low cooling load, only the air volume flow that is actually required for ambient

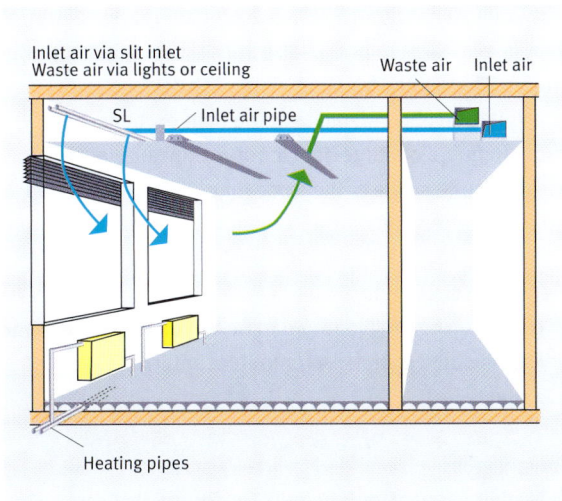

Fig. 3–40 Constant Volume Flow system (CVF)

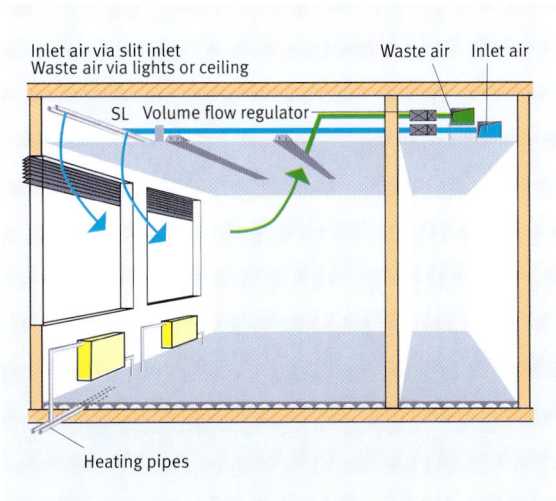

Fig. 3–41 Variable Volume Flow system (VVF)

air supply flows across the volume flow regulator. As cooling load in the room rises, e.g. through solar influence or communication devices, the volume flow regulator opens and allows the air volume required for cooling to pass through.

Air-Water systems

Air-Only systems distinguish themselves by the fact that thermal air processing takes place in the respective control center. Since, in comparison to water, air can transport a lot less energy at equal volume, the same performance level requires only a small volume of water to be transported. This is reflected by smaller canal pipes and pipelines but also by a lower volume of transport energy. In the case of Air-Water systems, practically each zone can be equipped with an after-treatment device and hence, an individual reaction to the respective heat loads can be achieved.

High Pressure Induction Air Conditioning systems:
In contrast to the air conditioning systems described above, space requirement for climate control centers, and ducts is considerably lower. One significant difference is that the entire primary air that is guided from the climate control center into the rooms is

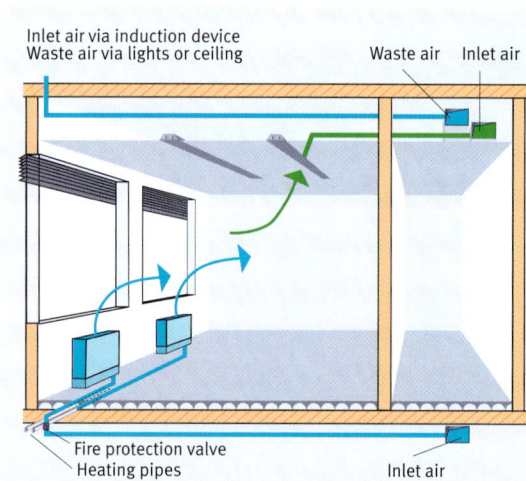

Fig. 3–42 High Pressure Induction Air Conditioning system

actually processed outside air, while for low pressure systems, for instance, only about 20 to 30% of outside air is taken in but the rest consists of circulating air.

The induction devices can either be installed in the ceiling void or in the parapet area.

Ventilator-Convector (Fan Coil) and CVF system:
In this system, ambient air supply with a 2 to 3-hourly air exchange is undertaken via a CVF system and (the largest part of) load dissipation is handled by the ventilator-convectors. These constitute de-central components, which are installed in the rooms to be conditioned. Situated inside a case with air suction and air blowout grid, we find a ventilator (often adjustable at three levels) and one or two heat exchangers for heating and cooling.

Indoor temperature adjustment is via water regulation valves and rpm control of the engine.
Fan Coils, as a rule, are equipped with a four-wire system, meaning that hot and cold water is supplied to the heat exchangers for heating and cooling in separate flow and return pipes. If applicable, an additional condensation pipe is to be planned for. Fan Coils and CVF systems are rarely installed in office buildings but frequently in hotels (fast conditioning). Fan Coils can also be run on pure ambient air operation, meaning without connection to a CVF system. In these cases, outside air supply is via the windows or via additional primary air systems. Fan Coils are frequently also used for computer control rooms, electric and phone distributor rooms and copy rooms. There, they function as cooling devices (without precipitation of condensate at the heat exchanger), in order to guide off large heat loads during ambient air operation mode, while at the same time, there is minimum space requirement.

General assessment:
– Maximum cooling load: 40 to 60 W/m² ventilated area
– For sensitive coolers in computer centers up to 1.000 W/m²

Advantages:
– Minimum space requirement for control devices

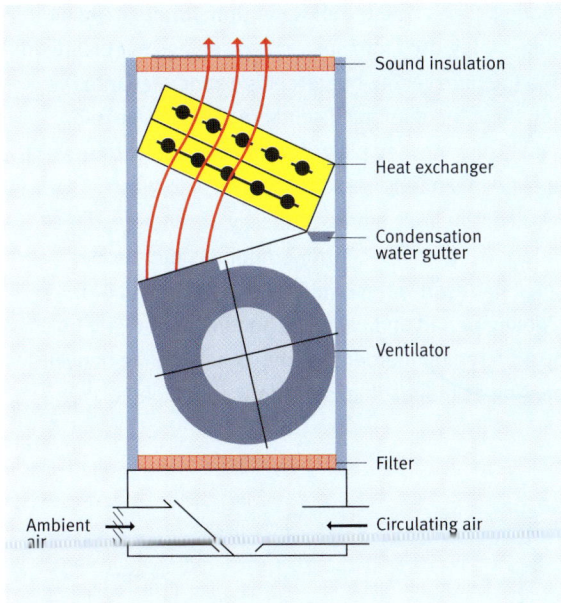

Fig. 3–43 Fan Coil system (5:6), Source LTG

General assessment:
– Maximum cooling load: 20 to 45 W/m² office area

Advantages:
– Low investment costs
– Absence of central devices
– No duct installation, resulting in low story height

Disadvantages:
– High maintenance costs
– Performance area is restricted

Cooling ceiling and CVF system: Depending on make, cooling ceilings consist of water through-flow cooling coils made of copper pipes or plastic pipe mats, which are integrated into a suspended ceiling structure. This ceiling structure may come as a stucco ceiling, metal panel ceiling or also as an open grid ceiling. Cooling ceilings can be designed as large-area ceilings but also as "islands" directly situated above the working area.

Facade ventilation devices: The disadvantage of the Fan Coil, which is only running on circulating air mode, has been absorbed through development of facade ventilation systems. Principally, facade ventilation systems can be imagined as small box devices that are arranged along the facade. They serve for climate control of smaller rooms, preferably office rooms. Since these devices ought to be as small as possible, the individual components cannot be arranged according to the modular design principle. Arrangement of the individual components is subject to certain additional boundary conditions that to not apply for central ventilation devices.

Pure inlet air devices are generally located in the hollow floor. They are also called underfloor systems. To increase performance, circulation air from the room can be induced in some devices.

Combined inlet and waste air systems are integrated into the facade in the parapet region. They are fundamentally also equipped with heat recovery. For all facade ventilation systems, care must be taken that the respective components enable them to compensate for wind pressure.

A general advantage of cooling ceilings lies in the fact that about half of the cooling load is discharged via heat absorption and the other half via convection. The absorbing effect of the cooling ceiling has an essential influence on well-being, if there is a small temperature difference between cooling ceiling and body it is perceived as comfortable.

Convective heat transfer, which due to high air volume is often accompanied by draught, is significantly less compared to Air-Only systems. Another advantage is that indoor air temperature for cooling ceiling systems, at equal operative temperature, can be about 1 to 2 °C higher than is the case for exclusively mechanical ventilation.

Regulation of indoor temperature is by water, for instance via simple thermostat valves or via sequence control to the radiator. To prevent dew water formation, and in the case of windows that can be opened, cold water temperature ought to be regulated in a sliding manner room-by-room and depending on indoor air dew point.

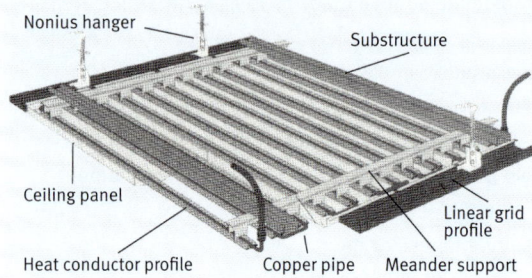

Fig. 3–44 Example for a cooling ceiling (5:10), Source Zent-Frenger

cooling. Pipes where cool water flows through are no longer integrated into suspended ceilings nor plastered into the bare ceiling but rather are concreted into the middle of the reinforced concrete ceiling. As a rule, quality intermeshed plastic pipes are used for this, which are laid meander-type at spacing of 15 to 30 cm. As a result of the huge storage capacity of concrete, direct indoor temperature control is not possible. Hence, constant mean water temperatures between 20 °C (cooling period and ca. 26 °C heating period) are to be adhered to in order to avoid under-cooling or over-heating as the result of short-term room loads.

This system is energetically especially cost-effective, since temperature level of cold water lies between 17 to 19 °C and not, like for air ventilation, at 6 °C flow temperature. This opens possibilities to undertake cold water generation without mechanical cooling generation, which is something we are going to discuss in more detail at a later stage

General assessment:
– Maximum cooling load: 80–120 W/m²
  Ceiling area = ca. 55–100 W/m² office area

Advantages:
– Very good thermal comfort level
– Low space requirement for technical installation
– Low energy costs
– Low maintenance costs
– Variable cooling performance
– Easy indoor temperature control

Disadvantages:
– High investment costs
– Restriction on architectural ceiling design
– Monitoring of dew point temperature for
  avoidance of condensate formation

Component cooling and CVF system: While cooling ceiling systems might be ideal from the view of thermal comfort levels, costs for this are very high, however. A much less expensive possibility of comfortable room cooling has been, for several years now, component

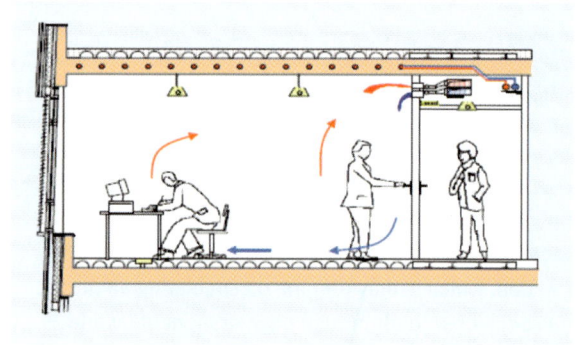

Fig. 3–45 Component cooling and heating

Owing to the largely constant surface temperature of the reinforced steel ceiling, radiation exchange with the environment results in a kind of autonomous regulation of indoor temperature. Cooling performance thus achieved – depending on piping, flow temperature and floor makeup – is ca. 20–40 W/m² of active ceiling surface. While this might be considerably lower than for the customary cooling ceilings, an advantage however is in the full-area effect of the total ceiling surface. Structurally caused taboo zones (e.g. In the area of shear reinforcement for flat ceilings) are to be considered.

Outer air supply is also via a CVF system, which can cover the ventilation heat demand. If one further takes into consideration the boundary conditions mentioned for the heating/cooling ceiling, one can also forego the usual heating system in front of the facade when it comes to component cooling/heating. However, mean heat transfer coefficient for the facade, in this case, must be at about ≤ 0,80 W/m² C, since heating performance of the system is significantly less than is the case for heating ceilings, as a result of the low excess temperature of mean water temperature versus indoor temperature (≤ 30 W/m²).

Fig. 3–46 Concrete core activation: Flex pipes embedded in concrete

As a result of its good cost-performance ratio, this system is nowadays increasingly used. Generally, the laying of the pipes is undertaken on mandate of the shell construction contractor. To avoid damage to the pipes, special care is required. Component cooling/ heating can only be achieved for rooms where no ceiling cladding is required, for instance for room acoustic or installation-engineering. Ceiling cladding prevents radiation exchange with the room-sealing areas and would significantly restrict component cooling/heating in its effectiveness.

General assessment:
– Cooling load: 10–50 W/m² office area

Advantages:
– Very good thermal comfort level
– Low space requirement for technical installation
– Low control engineering effort for one-zone system
– Low energy costs
– Low maintenance costs

Disadvantages:
– Restriction on architectural ceiling design
– No single room control of temperature possible

Waste air: Waste air openings in the room ought to be arranged as much as possible either on the ceiling or in ceiling vicinity in order to make sure that smoke can be safely suctioned out. From the exterior zone section, sucking off must be undertaken at a distance of 5 to 6 m from the window front.

For interior blinds use, a small amount of air is to be sucked off at the ceiling above the windowsill. Formation of indoor air openings provides considerable leeway without actually influencing their functioning: It is desirable in any event to guide off the air via lighting since, on one hand, even distribution of air openings is ensured and on the other a large proportion of heat generated by the lighting sources can be discharged directly without placing any stress on the room itself. This also reduces required cooling load of the overall system.

Waste air ducts for the individual stories need to be secured by fire dampers at the connection point to the main duct.

Ventilation and air conditioning control rooms: The dimension of a climate control room for a building depends greatly on the air volume required for acclimatization of the building. Total length is around 8 to 20 m, whereby air volume and required equipment with the different aggregates plays a decisive role for air processing. Width lies between 3 and 5 m. It is oriented on air volume to be transported and also on room height. Additionally, one service aisle in front of the climate control room, of a width between 1,5 m to 2,5 m is to be provided for in spatial planning.

An equipment room for small facilities ought to be at least 2,5 m high. For medium size and large facilities, room heights are common of up to 4 m. Floor loading in climate control rooms is about 1000 to 1500 kp/m². In this, weights of special footings for ventilators and pumps are already considered. Not included, on the other hand, are the weights of stonewalled or concreted

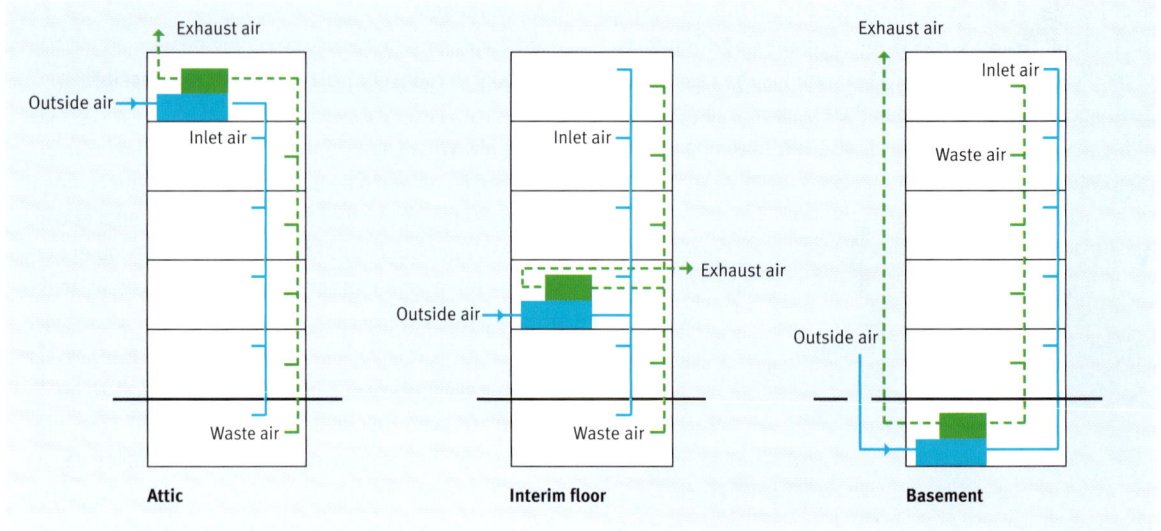

Fig. 3–47 Design options for VAC control rooms

air chambers as well as the concrete slab in case of a floating screed.

Arrangement of the control room: Ventilation and air conditioning technology control rooms can be arranged in the attic, in an intermediate story (frequent in high rises) or in a basement. Often, they are placed in the basements (rooms without daylight). Roof control rooms, generally, have the advantage that outside air and waste air do not need to be guided through the entire building. This means that elaborate outside air and waste air building and ducts can be avoided.

Design of the control room: In the past, components were mostly arranged in stonewalled or concreted control rooms.

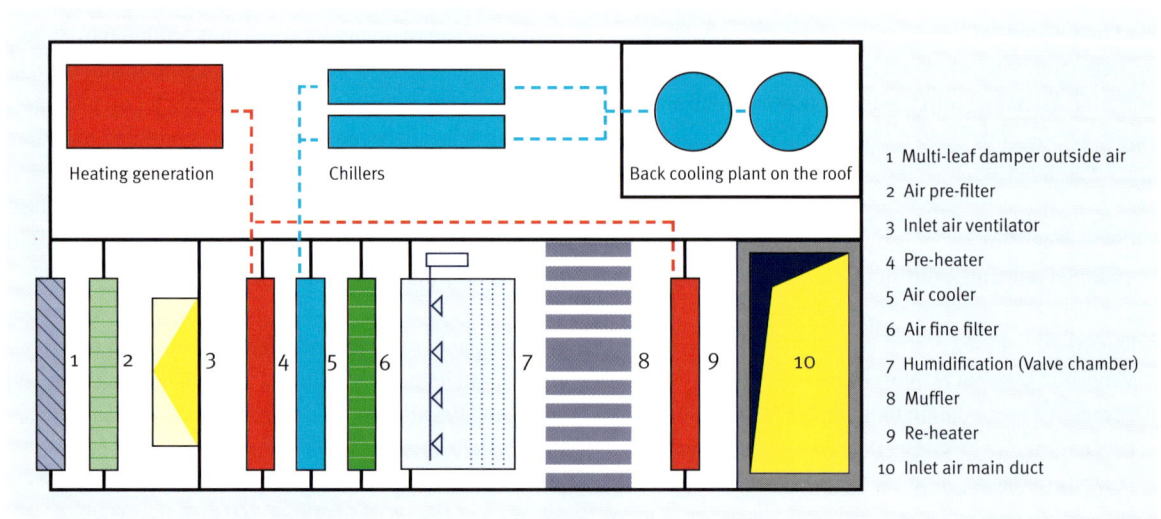

Fig. 3–48 Principle behind a VAC control room

1 Multi-leaf damper outside air
2 Air pre-filter
3 Inlet air ventilator
4 Pre-heater
5 Air cooler
6 Air fine filter
7 Humidification (Valve chamber)
8 Muffler
9 Re-heater
10 Inlet air main duct

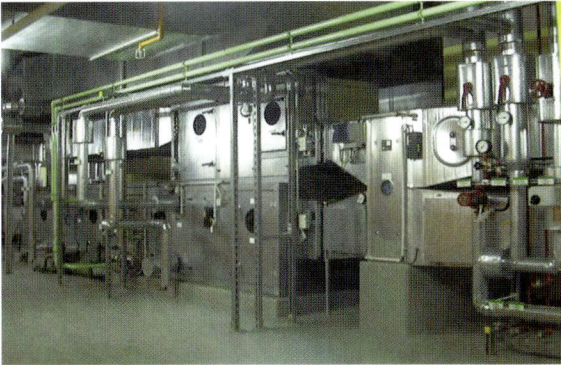

Fig. 3–49 Conventional climate control room

Nowadays, as a rule, box devices are put together from the various assembly groups, depending on requirement. The casing of the box devices consists of a profiled steel frame and a two-ply galvanized metal sheet mounted on top, with mineral fiber wool in between for heat and sound insulation.

Fig. 3–50 VAC device in box construction (Source GEA)

Doors are fitted for the purposes of inspection and maintenance. Into these devices, the respective required components are inserted like, for instance:

– Ventilators
– Heat exchanger (Heating, cooling)
– Filter
– Humidifier and mixing chambers
– Heat recovery unit and mufflers

In this, timely installation of the devices is to be considered.

3.4.5 Cooling Systems

Cooling systems (CS) are facilities that use certain agents (cooling agents) to withdraw heat from a given room and/or to cool it.

The CSs work with the cooling agent in a closed cycle. The CS changes its aggregate state during through-flow, since on one hand it draws out heat from its environment and evaporates it. On the other, the emission of heat causes the cooling agent to liquefy again. This process is known as a cycle.

For cooling generation, the two following processes are required:

The **cold vapor compression process** is the most widely spread since it is driven by an electrically actuated compressor. During this process, the cooling agent undergoes

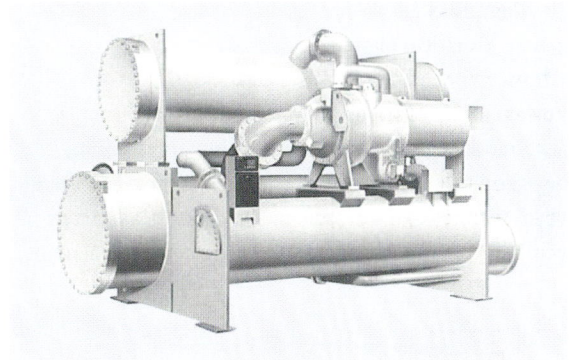

Fig. 3–51 Compression chiller with turbo compressor

gas-liquid states, where heat is taken up and emitted again. Example: fridge.

The **cold vapor absorption process** only works with an acting agent pair (cooling agent and solvent). Essentially, this process is driven by heating energy. The cycle of the cooling agent with the gas-liquid states is the same as for the cold vapor process, only with the difference that the solvent in the "thermal compactor" is initially added to the cooling agent (absorption) and then separated from it again (driven out).

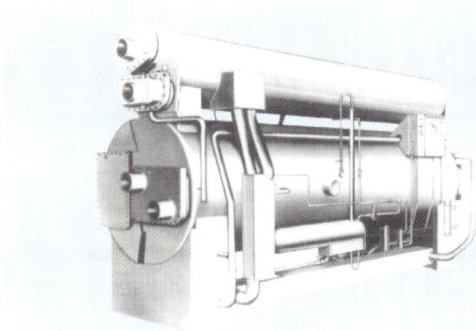

Fig. 3–52 Absorption Chiller

For this, a heating temperature of › 90 °C is required. Further, this process for cold generation in air conditioning systems runs in a vacuum. Therefore, a minimal amount of electricity is required for the vacuum pump and the process of solvent circulation.

3.4.6   Electrical Systems

Power current systems: In essence, power current is required for lighting and the operation of machinery/devices and/or servomotors. Major consumers are all conveyor systems like elevators and escalators as well as ventilators for air transport and large computer facilities.

From the main transformer station of the EVU, electricity is supplied – at a high voltage of up to 30 kV – to the transfer station. From the transfer station, electricity is then passed on to the load substations. These substations must be located in the load centers (For the other control rooms or the elevators, this means in separate building parts), since low voltage transmission across distances of over 60m is already not economical.

At the load centers, voltage is transformed to 230/400 V and the electricity supplied via the Low Voltage Switchgear (LVS) to the Floor Distributors (FD) or directly to the Major Consumers (MC). From the floor distributors, the electricity is passed on to the individual outlet groups.

Reading devices (Meters, rate control switches etc.) are on the medium voltage side for major consumers; in the case of several consumers, only low voltage reading can be considered.

In essence, the control rooms consists of three spatially separated areas:

Transformer with direct outside air connection and an interchange shaft, hence always located at the outer building edge.

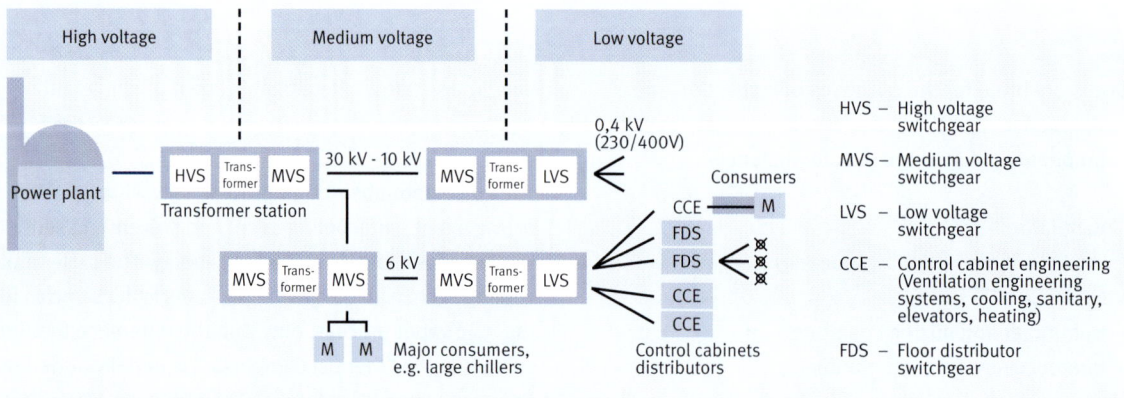

Fig. 3–53 High voltage switchgear floor distributor

Oil transformer · Cast resin transformer · Low voltage main distributor · Network replacement aggregate with cooling water cycle (green)

Fig. 3–54 Aggregates for power supply

Main strands: For major construction undertakings, the main strands nowadays are designed mainly as extended power rail systems, with the big advantage of right angle exits and high planning flexibility level. If cables are used for the main strands, a bending radius of up to 100 cm needs to be taken into account. The cables are less flexible and harder to lay but make up for it by being less pricey.

Equipment supply: For electricity supply of office equipment, desk lights, cleaning devices etc., a number of different systems have developed – especially in administration construction – that also serve for low voltage power supply.

Wall installation only suffices anymore for very simple buildings with a low degree of technology and most often is actually used for residential construction.

Wall trunking installation is widely spread for cubicle offices and offers sufficient options for simple to medium equipment furnishings. Wall trunking is easier to install and not expensive; there is a certain problem in sound carriage to the adjacent rooms, however.

If flexibility in room arrangement is required, this also requires a wide-area electricity supply. A cost-efficient but very rigid system is the **electrical duct in the raw ceiling**.

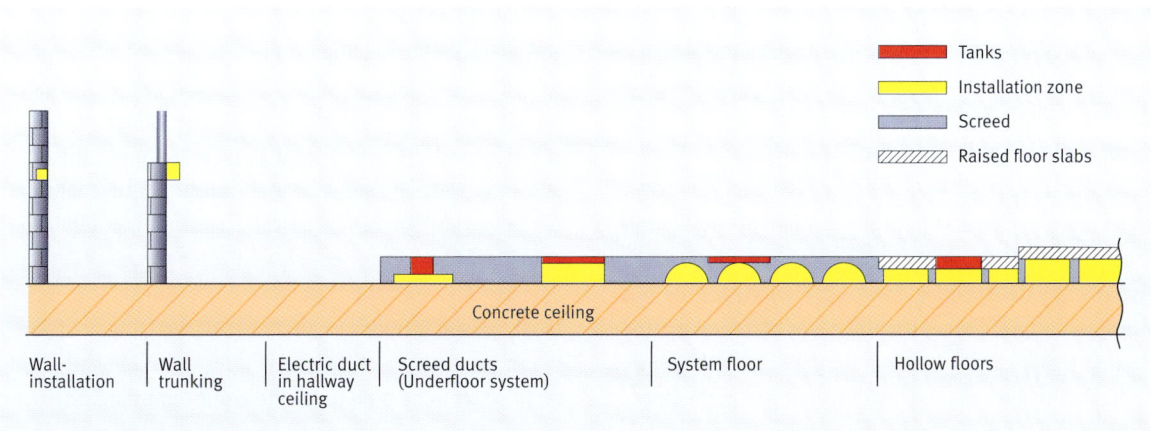

Tanks · Installation zone · Screed · Raised floor slabs

Concrete ceiling

Wall-installation | Wall trunking | Electric duct in hallway ceiling | Screed ducts (Underfloor system) | System floor | Hollow floors

Fig. 3–55 Different varieties of electrical installation

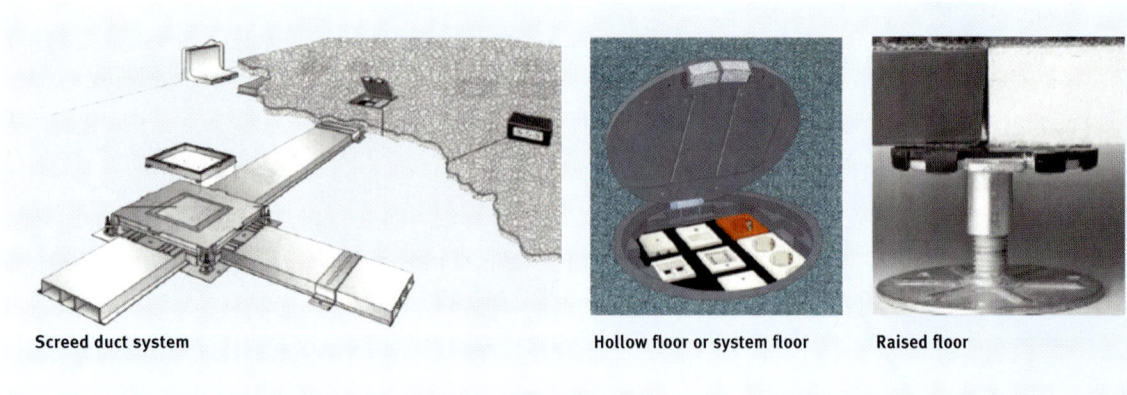

| Screed duct system | Hollow floor or system floor | Raised floor |

Fig. 3–56 Examples for floor installation

The ducts are inserted in between the reinforcement systems and once there, their location cannot be changed anymore.

Screed duct systems are also rather rigid, for the duct and exit positions need to be specified in planning early on. Change in case of emergency is possible but always connected to screed damage or destruction.

Of some flexibility, however, is the **hollow floor or system floor** approach, where a level line is cast across formwork resembling an egg carton – this results in a grid-like installation zone of 5 to 9 cm clearance height. While the junction boxes also need to be fixed prior to installation, upgrading through simple drilling open with a special drilling device is possible.

A fully flexible system, however, is the **raised floor** approach, which has been known for a long time. Here, individual slabs are stacked to whatever height is desired, resulting in an optimum installation zone.

Upgrade is possible at any time through taking in individual slabs. However, there are some acoustic problems, especially when it comes to wood planks, plus the floor fabric eventually loosens in the event of frequent subsequent upgrades. A problem that affects design more than anything else is the resulting pattern in the floor covering, which goes according to the slabs and only allows for certain qualities and patterns.

Overall, costs rise with increasing flexibility.

3.4.7   Weak Current Systems

Equipment supplied by weak current (12 V) in essence includes:

– Telecommunications systems
– Data transmission networks
– Intercom systems
– Antennae facilities
– Clock facilties
– Paging systems
– Electro-acoustic systems
– Alarm systems
– Video systems
– Access control facilities

Installation of the cables and wires, in essence, is undertaken as part of the high voltage installations in the same installation zones, whereby appropriate shielding of the low voltage cables must be absolutely taken into consideration.

Weak current facilities interfere heavily with organization planning – hence, decisions need to be made early on in order to be able to lay the empty conduit system in good time. This, especially, goes for optical signaling systems, telecommunications and call-by-call systems.

Together with the planning application, the applications for telecommunications systems ought to be submitted at the same time, since delivery time, depending on system size, can run up to two years. The same goes for the fire alarm system since it runs on the same lines.

Delivery time for a central control room is a minimum of nine months. Spatial requirements correspond to those of an EDP room (dust-free). Installation is according to instrumentation and control engineering of the in-house technical control rooms. Installation time is about four weeks. For this, the wiring diagrams and terminal schemes of the various firms are required.

3.4.8   Lighting

The manner of lighting orients itself on the respective requirements for the individual utilization areas. In the field of office buildings, for instance, we distinguish according to the following areas subject to different lighting:

– Working place lighting
– Lighting for meeting rooms
– Lighting for entrance halls and counter halls
– Lighting for circulation areas
– Lighting for adjacent rooms

In the case of working place lighting, a functional lighting approach is the prime consideration while, for meeting rooms, entrance and counter halls, a representative aspect is something that we place more emphasis on. For this reason, we tend to speak of artificial lighting technology for the working place area and for specialist areas we tend to use the term lighting design.

In office building construction, working place lighting is the prime consideration and we have various options for this.

The type of lighting that is still most widely spread today is ceiling lighting. If this is being used as general lighting, it is very flexible but not necessarily the optimum solution from the viewpoint of the individual workstations.

However, if this type of lighting is specifically adapted to the individual workstations then flexibility is lost. For this reason, desk and upright lights tend to become more and more popular, which are independent of the ceiling, since they are both workstation adapted and flexible at the same time. However, light exploitation from the ceiling systems is almost always better and energy consumption, hence, is less. However, with today's lighting and lamp technology that we have available to us, this is no longer such a grave factor.

In terms of the construction schedule, the entire lighting system is rather to be considered as a furnishing element that, with exception of integrated ceiling systems, is not added until the end.

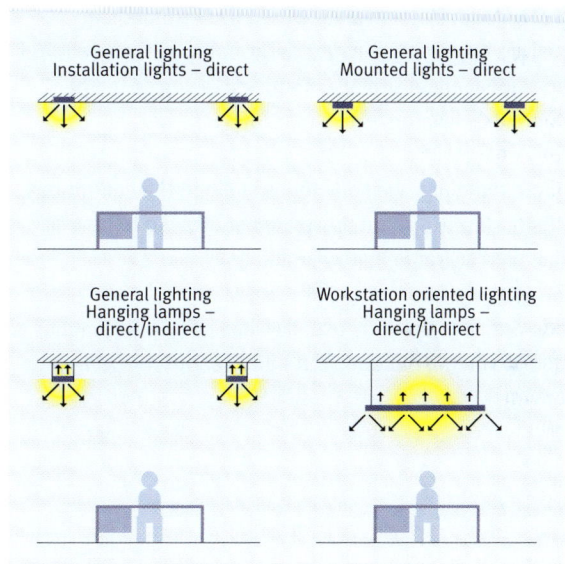

Fig. 3–57 Alternative ceiling lighting

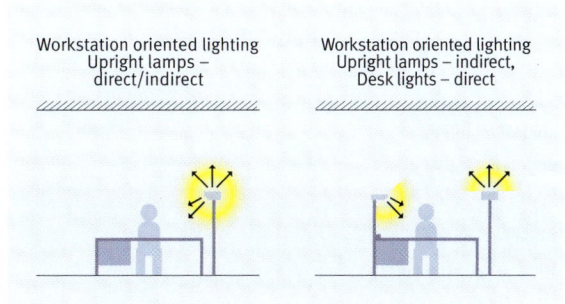

Fig. 3–58 Lighting that is independent of the ceiling

3.4.9 Conveyor Systems

Elevators: For the vertical transport of loads, traction-driven elevators are the usual choice. A special advantage is the high transport speed of 0,8 (alternating current drive) up to 2,5 m/sec (Direct current and/or linear drive).

The engine room is one story above the last floor that can be accessed with the elevator.

In the event of minimum height differences and high loads, a hydraulic elevator is often used. In this event, the last accessible floor does not require an engine room but additional foundation depth is required for accommodating the hydraulic piston.

If this depth is not available then two symmetrical pistons can be built in next to the cabin but this requires additional areas and also finances. Lifting speed of hydraulic elevators is at a max. of 0,8 m/sec.

In old buildings, we sometimes still find the continuous lifts (paternoster) but these are not allowed to be installed anymore in new buildings.

Construction elevators usually came as facilities mounted outside on the building as a barred cabin with gear rack drive.

Fig. 3–60 Example for a panoramic elevator

Awarding of this contract must go hand in hand with shell construction awarding or must happen prior to that, in order to ensure that the definite dimensions of the cores can be specified. One elevator ought to be provided as construction elevator (if possible, this should become the later freight elevator). Use as construction elevator must be recorded in the BOQ (Earlier TÜV inspection, cabin lining). The elevator attendant's salary ought to be included in the cost estimate, since passing this on to the respective firms is rather difficult. If passing on of these costs is intended then this must be included in all BOQs (% specification).

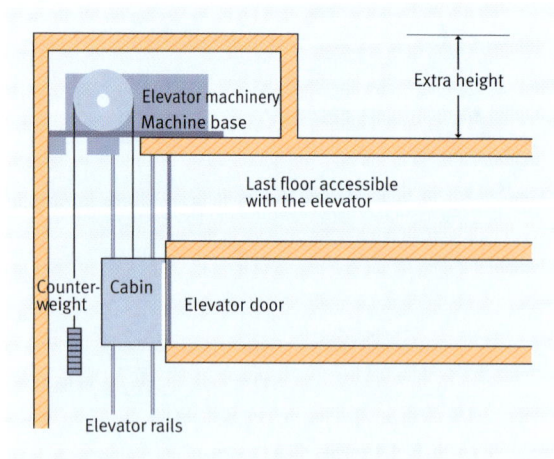

Fig. 3–59 Principle of the traction elevator

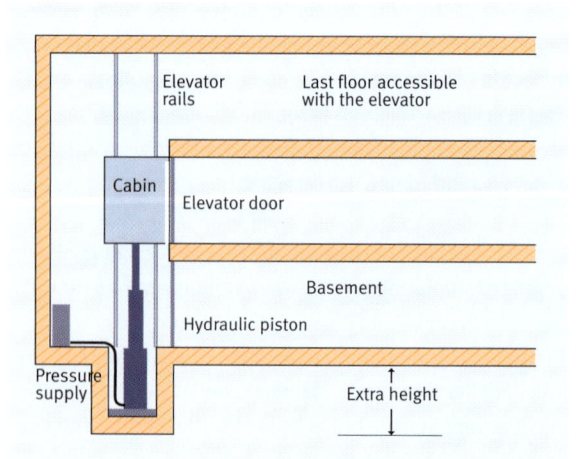

Fig. 3–61 Principle of a hydraulic elevator

Escalators: Escalator facilities consist of a drive motor, lateral guiding rails where the escalator segments run, and the side rail with hand protection. Escalators are usually factory-assembled and then built on location as a one-piece unit.

A moving walkway is one variety of the escalator where the individual segments do not run horizontal but according to the rising tilt. Construction length for moving walkways is much larger than for escalators.

Awarding of this contract must go hand in hand with shell construction awarding or must happen prior to that, since it has an impact on formwork and reinforcement plans, sometimes also on static equilibrium.

These options exist for installation of these facilities:

– Insertion from above prior to closing of the roof (Little danger of damage, only if buildings are low due to the required lifting tools)
– Insertion from above after each ceiling (Danger of damage, frequent use of lifting tools)
– Lateral insertion after roof has been closed (Facade must remain open in the region of the installation opening; the entire transport route to the insertion site cannot be taken down)

High-rise elevators: Essential transport routes in high-rises must be negotiated via elevator.

Depending on the guidelines, a max. of three elevators is permissible per shaft. The frontal zone ought to have 1,5 – 2 x cabin depth and be free of crossings. From four adjacent elevators already, clearness aspect is lost and required entry and exit times become too long. Generally, mean waiting times of 10 seconds are considered to be very good, 15 seconds good and 20 seconds as just about sufficient still.

Elevators ought to be designed for a maximum people load of 20 since, otherwise, too much time is required for entry and exit. To grant adherence to a "personal space" minimum or at least not to infringe on this personal space too much, generous design is sensible. While panoramic elevators offer more space optically, they are not suitable for all passengers.

Depending on comfort level, use and building size, there are many different access principles for elevator facilities.

Individual access: Each floor can be reached without the need to change, whereby, initially, express local elevators can be used. Accessibility to only ca. 40 floors makes sense since, otherwise, too much space is required in the plinth region. Unused remaining spaces in the upper floors can be used for other applications.

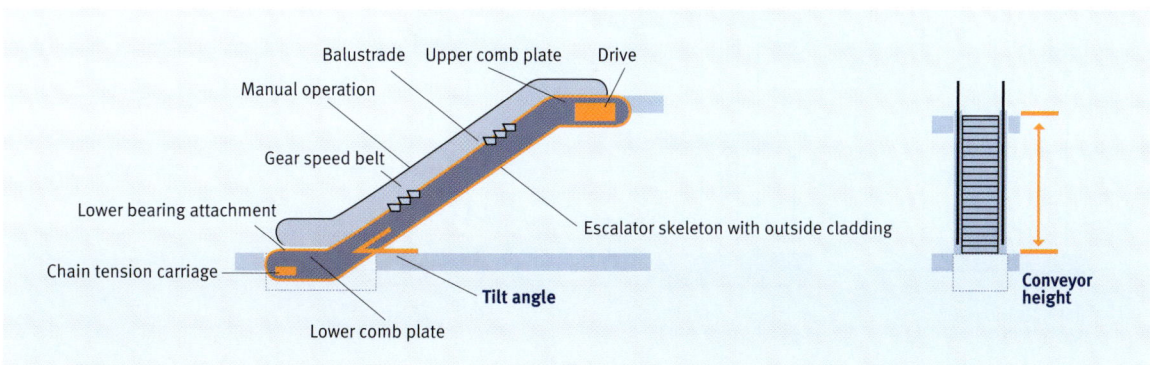

Fig. 3–62 Principle behind an escalator system

Progressive individual access: Allows space saving through stacking the elevators on top of each other, yet it requires changing elevators.

Block access: Express and local elevators separated from each other also lead to a space-conserving arrangement. Further, here and for all other accessibility principles, there is the requirement of arranging for overpass and underpass between the local elevators. The non-accessible floor can then be planned as a technology floor.

Block access with escalators: For handling large circulation volumes, access or subdivision on the lower levels can be handled via escalators.

Freight and fire fighter elevators: Aside from elevators serving for access purposes, an arrangement of freight and fire fighter elevators is also required. These need to be situated in a separate shaft and be equipped with an emergency power generator.

Design of vertical access: Analogue to experiencing a city, the adventure of vertical access through correspondingly open and hence public access elements is possible. Not only the transport process itself but also the creation of new experiences can be the aim here, whereby open and inviting stairs can constitute socialization and movement friendly routes while escalators and panoramic elevators ensure comfortable experience of vertical spatial movement at great heights. As an architectural design expression and at the same time as a statement for an active society, vertical access elements on the inside and outside of future high-rises can become an adventure potential that has, thus far, received little consideration and is purely high-rise specific.

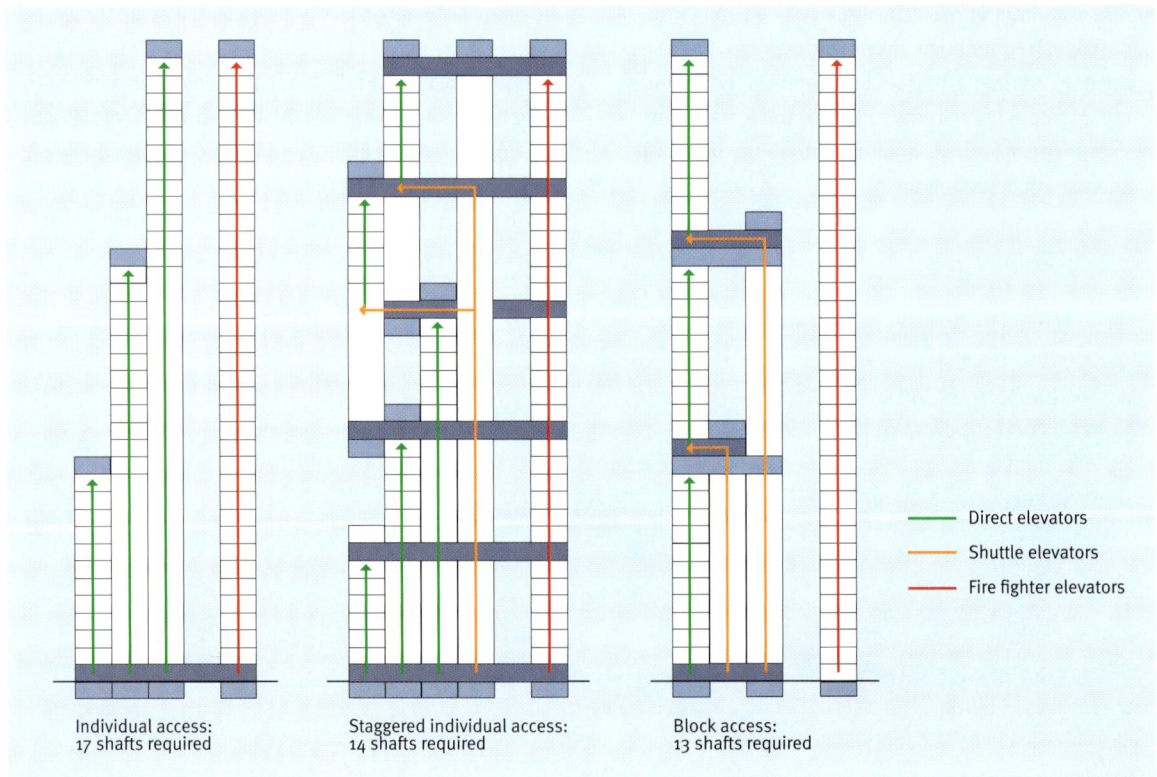

Individual access:
17 shafts required

Staggered individual access:
14 shafts required

Block access:
13 shafts required

— Direct elevators
— Shuttle elevators
— Fire fighter elevators

Fig. 3–63 Alternative elevator access in high-rises

3.4.10 Building Automation

The least complex facilities are simple malfunction processing centers where the malfunction of technical equipment is indicated via malfunction alarm lights. More comfortable malfunction processing centers use the respective wire diagram for closer specification of the malfunction originating from the control room.

Our modern and complex buildings would nowadays be unthinkable without building services equipment. These technological systems require a superseding system that automatically controls and regulates the individual system facility but also interconnects the various facilities and hence ensures requirements to be met of individual system facilities on other facilities.

This superseding system is what we know as building automation with its underlying measurement and control technology. Here, intelligent and de-centrally arranged sensors, actuators and controllers (the so-called components) take over control of any and all trades present in the building like, for instance, heating, ventilation, cooling, electrical systems, lighting, control of the blinds, shading, access control, monitoring systems, safety and energy management. The individual automation steps are known as a process.

Aim of building automation:

– Control and regulation, according to demand and on the basis of consumption optimization, of the Building Services Engineering (BSE) facilities
– To support the respective staff in their tasks of troubleshooting, operation, optimization and energy consumption monitoring.
– To pass on malfunction alarms and general alarms to the correct site
– To enable the gathering of several buildings into one operation control room
– To increase overall availability of engineering systems through registration of all malfunction alarms
– To achieve savings through adjusted energy management programs

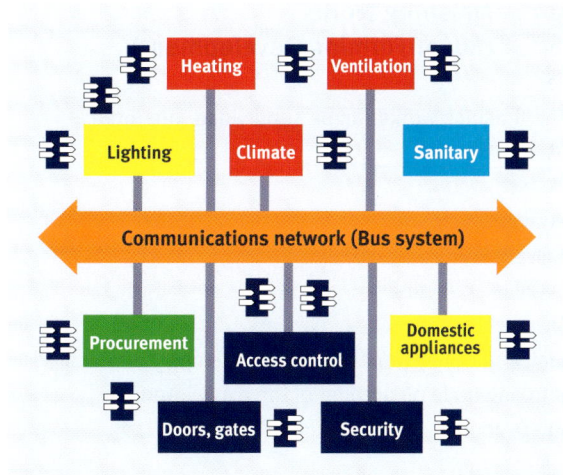

Fig. 3–64 Building automation

3.5 Finishing Works (Room-Forming Development)

The finishing works can be subdivided into three essential areas:

– Floor coverings
– Walls and wall paneling
– Ceiling paneling

Wall and ceiling paneling are increasingly being united into one trade on account of the many interfaces.

3.5.1 Floor Structures

Floor construction in administrative buildings depends very strongly on ground plan planning, available story height, economical considerations desired flexibility.

The classic raised floor systems are the most flexible but also the most expensive. Depending on flexibility require-ments, screed systems with either cable ducts or hollow floors are more favorable. The less certain the final floor plan is, the more economical hollow floors get.

The simplest and cheapest versions are screeds with and without impact sound absorption and without installation possibilities.

Fig. 3–66 Alternative floor coverings (Carpet, parquet, natural stone)

Any top floor covering in principle, can be put on the respective floor construction. Difficulties are usually encountered with stone coatings on access floors, though. This combination should be avoided if possible or only executed together with cable ducts and defined connections.

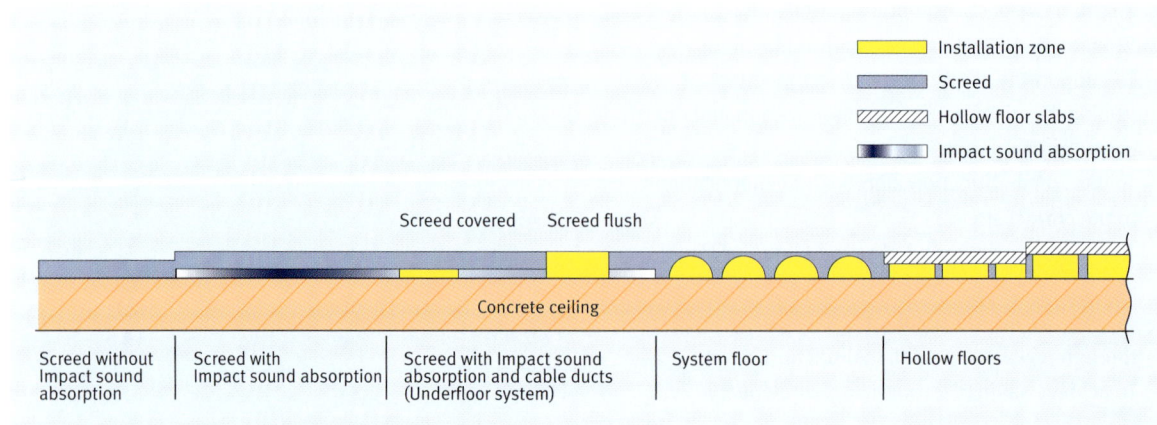

Installation zone
Screed
Hollow floor slabs
Impact sound absorption

Screed covered   Screed flush

Concrete ceiling

| Screed without Impact sound absorption | Screed with Impact sound absorption | Screed with Impact sound absorption and cable ducts (Underfloor system) | System floor | Hollow floors |

Fig. 3–65 Alternative floor systems

3.5.2 Partition Wall Systems

Light assembly design dominates nowadays as partition walls in office building construction but increasingly also in high-quality residential construction.

Heavy partition walls made of masonry or concrete are only used as partition walls of rental units, as the end partition to circulation areas or for fire protection reasons. They are then often used as constructive walls for load transfer and reinforcement. Surface is usually plastered with pain or wallpaper.

The assembly walls consist of a light sub-construction made of wood or metal profiles and can achieve soundproofing of over 40 dB easily, with a corresponding construction. In addition, they are easy to install by electric lines being led through prior to closing the plating. The surface can be painted or wallpapered.

These systems are also described as "easily destructible" and they cannot be moved. For high flexibility requirements, however, there is precisely this demand on the partition walls. This is achieved by so-called element walls. They consist of fully prefabricated wall elements, which, as a rule, can be connected to suspended ceilings with corresponding detail formations.

To move it, you simply undo the locking, lower the wall and then install it at another place. As a rule, the surfaces come readily painted. Newer developments also allow for repainting or wallpapering. The prefabricated elements can also be used as built in wardrobes and, as system wall, also serve to close off the hallway.

Fig. 3–68 Example light partition wall

If rooms are to be sectionel off temporarily during later operation, mobile partition walls are required. These can be guided along the ceiling in a track system and stored in so-called wall packages.

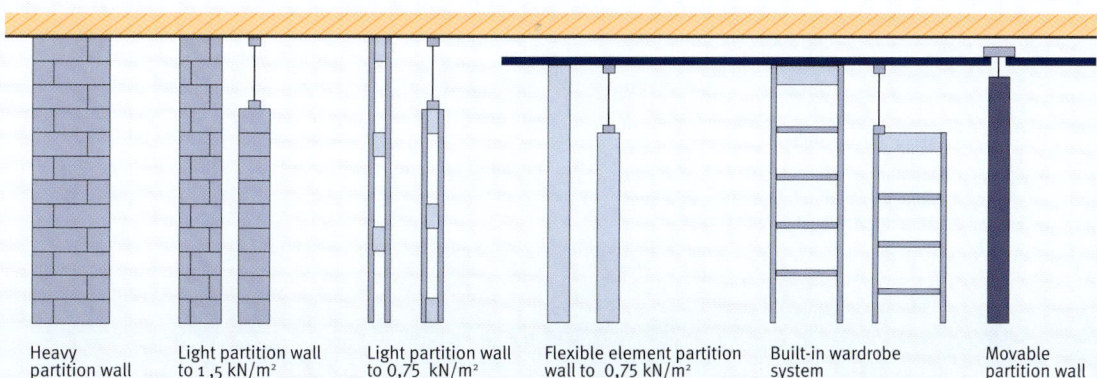

| Heavy partition wall | Light partition wall to 1,5 kN/m² | Light partition wall to 0,75 kN/m² | Flexible element partition wall to 0,75 kN/m² | Built-in wardrobe system | Movable partition wall |

Fig. 3–67 Alternative partition wall construction

3.5.3 Ceiling Systems

Although they are unfavorable from a building physics point of view, the majority of office buildings are equipped with suspended ceilings. They essentially serve for disguising of installations and to accommodate built-in ceiling lighting.

If the partition walls between the individual rooms go through to the raw ceiling, their installation is relatively simple. In this case, the roof structure needs only absorb its own weight and the lighting, possibly also ventilation outlets. If, however, flexible partition wall elements are to be attached to the ceiling in the suspended ceiling frame, it must be either designed as a sound insulation ceiling in order to stop the sound from traveling from room to room or bulkheads must be installed in every possible axis.

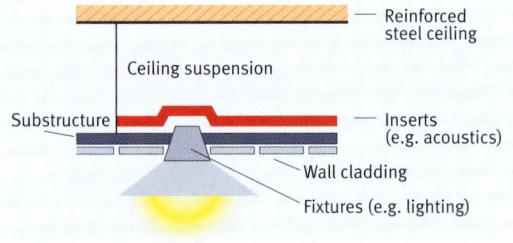

Fig. 3–69 Construction principles suspended ceiling

In the axis itself, a special detail for absorbing of the partition wall connection must be integrated or, if no partition wall is to stand in the axis, of a blind. In connection with a required axis-wise installation, this leads to high costs. The procedure for such a system is illustrated in the following graph.

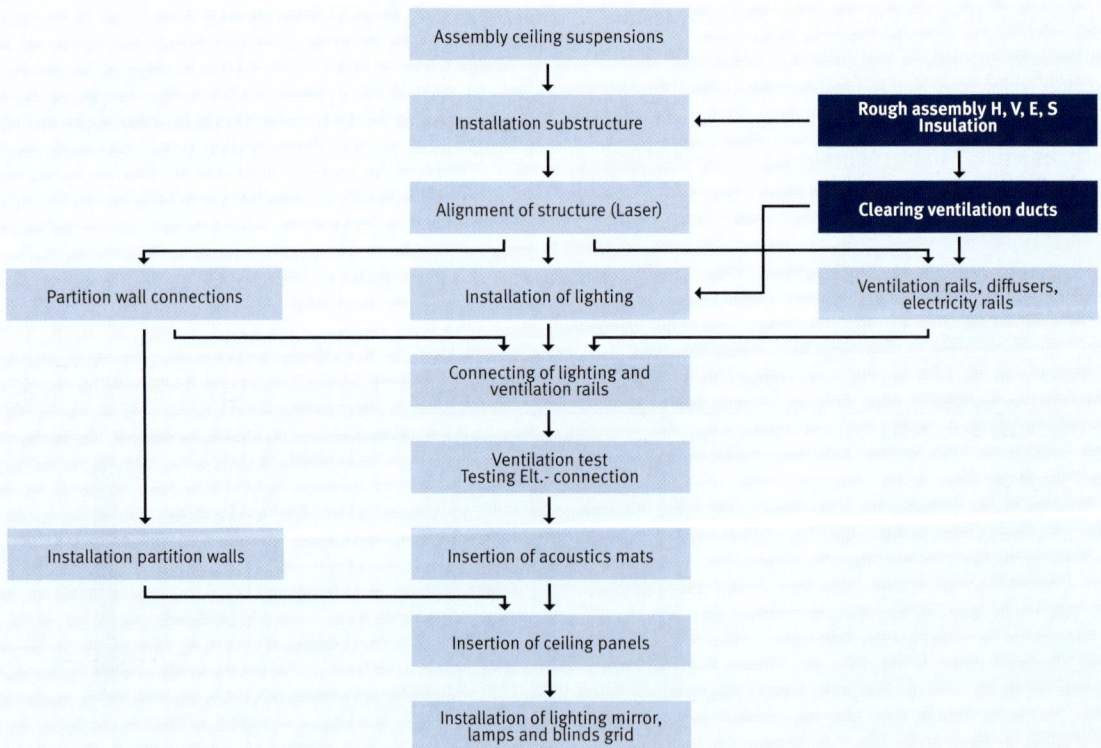

Fig. 3–70 Assembly process for suspended ceilings

In this process, one can identify the various dependences from the beginning of the ceiling suspensions to the installation of panels, light mirror and blind grid.

3.5.4 Process Variants – Finishing Works

The possible process for the finishing works depends very strongly on the systematic approach used for Floor-Wall-Ceiling. Some varieties are now shown at the example of a highly installed office zone with assembly walls made of gypsum (in the case of variant 3 assembly walls), a hollow floor and ceiling and floor installation.

Variant 1, as a rule, is only used for hallway floors or for community rooms, since otherwise too much effort would be required in case of modifications being required. With this variant, works need to commence either with installation of partition walls or with ceiling

installation or the suspended ceiling needs to be completed and afterwards the entire floor installed and arranged or vice versa. Process A is to be recommended, since the assembly walls are to be precisely aligned and the installation can be adjusted to this. Process C is also possible.

Since variant 2 has the partition walls actually standing on the floor, they must be installed and arranged before then. However, to avoid unnecessary soiling, the rough ceiling installation ought to be executed before then. Hence, process A is to be recommended. In any event, the suspended ceiling is to be installed last.

For variant 3, the partition wall comes at the end in any event. Once more, to avoid soiling and damaging of the floor, process A is to be recommended. Here, too, lighting equipment and ceiling panels are installed right at the end.

Variant 1: Partition wall from bare floor to bare ceiling (total sound insulation from partition wall)

1	Ceiling installation
2	Suspended ceiling
3	Hollow floor
4	Installation floor
5	Partition wall

Process:
A 5 → 1 → 2 → 4 → 3
B 1 → 5 → 2 → 4 → 3

Variant 2: Partition wall from finished floor to bare ceiling (sound insulation measures required for the floor)

1	Ceiling installation
2	Suspended ceiling
3	Hollow floor
4	Installation floor
5	Partition wall

Process:
A 1 → 4 → 3 → 5 → 2
B 4 → 3 → 5 → 1 → 2

Variant 3: Assembly partition wall from finished floor to suspended ceiling (sound insulation ceiling and measures required for the floor)

1	Ceiling installation
2	Suspended ceiling
3	Hollow floor
4	Installation floor
5	Partition wall

Process:
A 1 → 2 → 4 → 3 → 5
B 1 → 4 → 3 → 2 → 5

Fig. 3–71 Process varieties depending on partition wall structure

3.6   Consequences of sustainable Design

The construction and real estate industries touch on and unite many different sectors where sustainable thought and action continues to become increasingly more important – for market economy reasons as well as for environmental aspects. The idea is to carefully handle available resources and to apply regenerative energies and materials as much as possible. When it is implemented in an intelligent manner from a planning point of view, this idea is not only of ecological benefit but also contributes to adding value to the property while also reducing operating expenses. Staying within, or even below, legally specified requirements in the fields of energy efficiency, environmental protection and health protection also reduces the risk of loss of value and future upgrading expenses.

Starting points for improving the sustainability of buildings ought to always be considered from ecological, economical and social points of view. They can be divided into different categories:

3.6.1   Reduction of Energy Requirements and Use of Regenerative Energies

Owing to the worldwide technological advances and global population increase, an ever-larger number of humans keep requiring more and more energy resources. Connected with steadily rising prices for fossil fuels, efficient handling of energy, hence, is going to be required in future in order to construct or upgrade real estate in a marketable manner. Aside from active environmental protection, measures like efficient heat insulation and optimized heat-lighting distribution are decisive in determining economical aspects of a given building. In Germany, there is an obligatory energy pass since 2007 – as part of the energy saving ordinance, in German: Energiesparverordnung (EnEV) – and also, low primary energy requirement is actually a legal obligation. Primary energy requirement, used as a measured value, corresponds to the amount of energy required for the generation, transformation, distribution and transfer of energy consumed. The idea is to reduce this amount as much as possible.

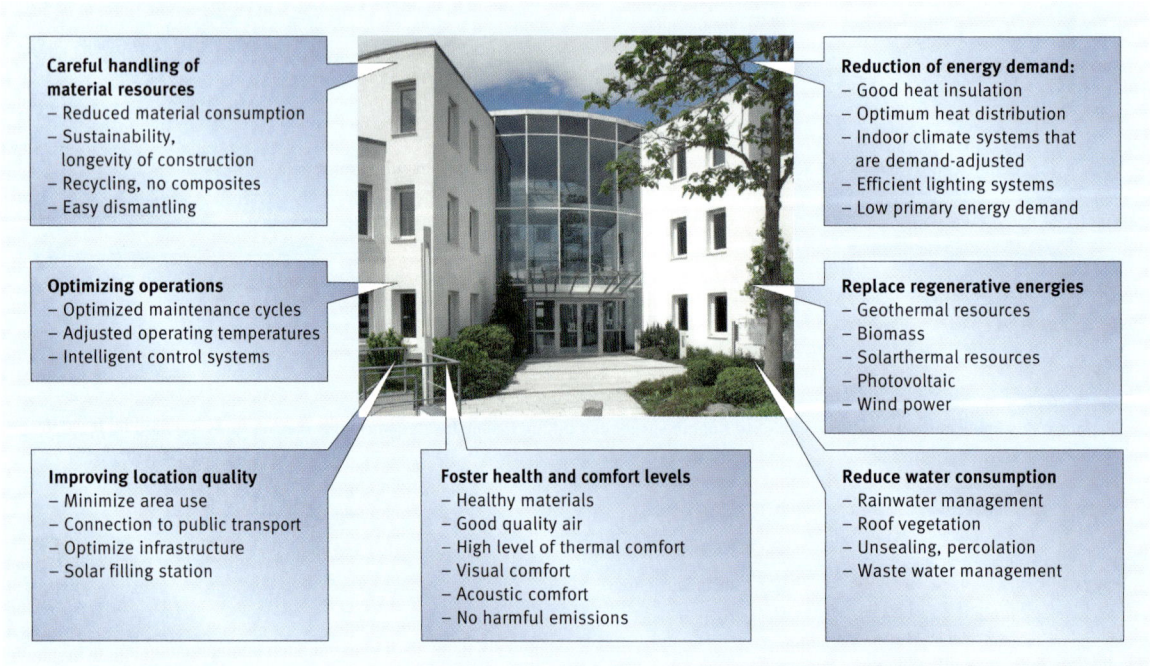

Careful handling of material resources
- Reduced material consumption
- Sustainability, longevity of construction
- Recycling, no composites
- Easy dismantling

Optimizing operations
- Optimized maintenance cycles
- Adjusted operating temperatures
- Intelligent control systems

Improving location quality
- Minimize area use
- Connection to public transport
- Optimize infrastructure
- Solar filling station

Foster health and comfort levels
- Healthy materials
- Good quality air
- High level of thermal comfort
- Visual comfort
- Acoustic comfort
- No harmful emissions

Reduction of energy demand:
- Good heat insulation
- Optimum heat distribution
- Indoor climate systems that are demand-adjusted
- Efficient lighting systems
- Low primary energy demand

Replace regenerative energies
- Geothermal resources
- Biomass
- Solarthermal resources
- Photovoltaic
- Wind power

Reduce water consumption
- Rainwater management
- Roof vegetation
- Unsealing, percolation
- Waste water management

Fig. 3–72 Sustainablity goals

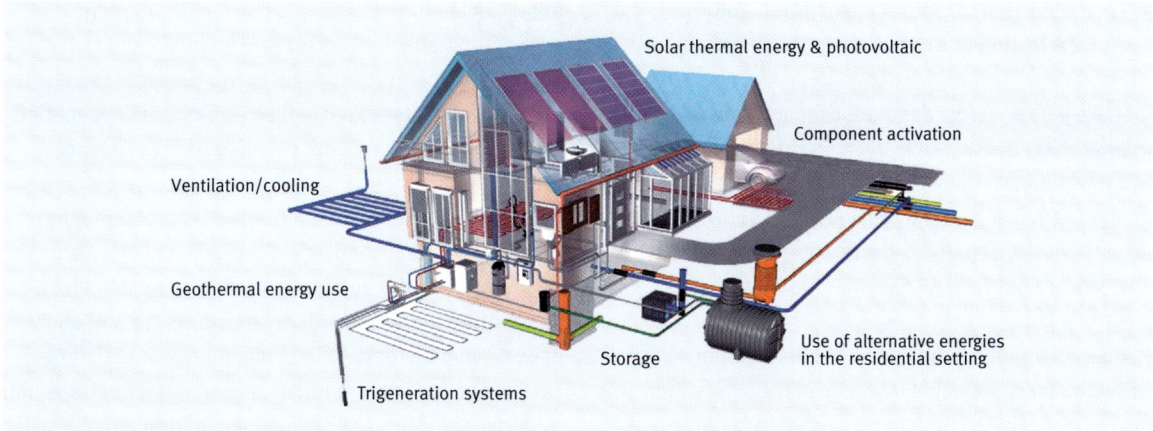

Fig. 3–73 Building blocks of ecological building conditioning (Source: BM Verkehr, Bau und Stadtentwicklung)

For high-rises, also, energy requirement can be significantly reduced through optimized heat Insulation in connection with use of passive solar energy. Hence, traditional air conditioning systems are not required; instead, cleverly thought-through facade concepts with natural ventilation are used – for high-rises as double-skin facade and box windows – as well as storage capacity of walls and ceilings on the interior of the building exploited. Small-dimension ventilation systems and cooling via cooling ceilings support a mainly naturally generated indoor climate in part-sections. The end effect is that, nowadays, over 50 % of energy is saved in comparison to traditional air conditioning systems that were used in the old days.

Over recent years, research in the field of renewable energies has made enormous progress so that resource-conserving energy generation and economical building operation are no longer mutually exclusive. In the contrary, due to favorable crude material prices and shorter transport routes, many renewable energy resources like wood pellets or geothermal heat meanwhile constitute a less pricey alternative to oil, coal or gas.

Additionally, primary energy of pellets is lower by 20 % in comparison to fossil fuels, something that bears enormous advantages especially in view of the EnEV. Further, development of environmental energies like wind power, photovoltaic or solar thermal energy can pay for off if applied correctly.

In the case of large inner city areas, trigeneration can serve to supply several buildings with electricity. Waste heat from electricity generation can then be used for heating or cooling over comparatively short transmission lines. These synergies reduce CO_2 emission to one quarter when compared to possible auto-supply with heating fuel or gas!

Combining a reduction of energy requirement with the use of regenerative energies is a core approach of sustainable construction.

The correct sequence of the process is decisive. This means, for project management, that there needs to be correct information about the possibilities that need to be defined and analyzed in the early planning stages.

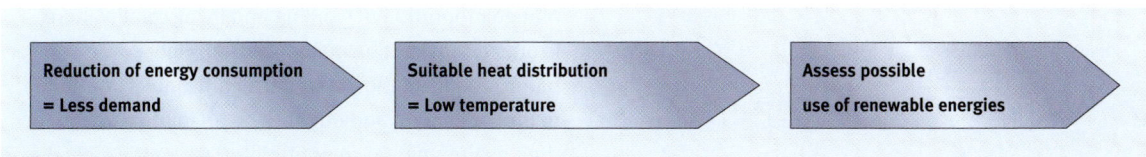

Reduction of energy consumption = Less demand	Suitable heat distribution = Low temperature	Assess possible use of renewable energies

Fig. 3–74 Three steps towards an energy-efficient building

3.6.2 Reduction of Water Consumption and Protection of Ground Water

Just like global energy demand, consumption of drinking water has also steadily been on the rise. Increased water withdrawal from the environment goes hand in hand with rising pollution levels. This shows in both water prices and costs for waste water disposal, both of which are particularly high in Germany.

Hence, care needs to be taken that the buildings are designed for sustainable handling of drinking water. This means, especially, careful approaches to lavatory use, personal hygiene and clothes washing. A more recent development in this is the use of "waterless toilets" as we know them from airplanes and high-speed trains.

Collection of rainwater, especially in conjunction with roof vegetation, and its use of processed industrial water or for irrigation purposes, saves costs in the medium term as well. Through unsealing paved surfaces etc., natural water cycles can be improved and ground-water generation fostered in a sustainable manner.

Example Potsdamer Platz in Berlin: A large proportion of rainwater is stored through plants and the roofs with vegetation. This stored water contributes to a positive

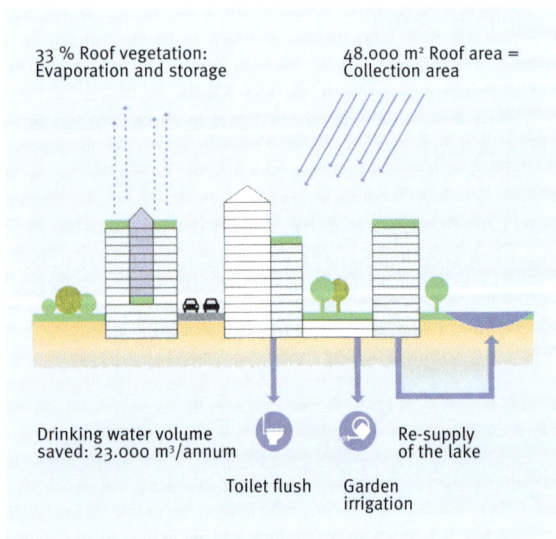

Fig. 3–75 Conservation of water resources

microclimate in the vicinity of the roof through the evaporation effect, by increasing humidity and reducing summer temperatures by one to two degrees. The remaining rainwater that is not absorbed by the vegetated roofs is collected and kept cool and dark in cisterns. 48.000 square meters of roof area supply a water volume of 23 million liters per annum, which is processed by means of simple technologies. One million liters are used for irrigation of the outside areas and 12 million liters for re-supply of an artificial lake. Afterwards, 10 million liters of water remain for toilet flushing for several buildings. Hence, not one cubic meter of unused rainwater flows into the sewage system, 23 million liters of drinking water are saved and the microclimate improved. Such approaches, also, need to be incorporated into the project early on.

3.6.3 Health, Comfort and environmentally-friendly Components

Environmental protection also means to protect the immediate environment of staff and occupants and to foster them in a health-advancing manner. Only that way, a comfortable and hence efficient working day is possible.

A human being residing in industrial nations, nowadays, spends some 85 % of his or her lifetime inside buildings. Hence, the climate, acoustics and air quality prevailing in a building as well as its interior design, have an immediate impact on the physical and psychological condition of its users. The advantages of an environment that is designed according to these requirements are clearly at hand: stress or health-related loss decrease, occupants identify more clearly with their work, are more motivated and are able to work in a more concentrated manner. This also decreases salary costs while at the same time increasing staff efficiency.

A huge proportion of working and living quality, here, is due to construction materials used: Floor coverings, paints and insulation material significantly contribute to indoor climate and ought to be absolutely tested for their environmental compatibility. Frequently, only the first heating period shows whether solvents or adhesives

Fig. 3–76 Positive socio-cultural working environment

emit climate-damaging and health-damaging gasses, which can also be cancerous in extreme cases. The influence on staff of significant odor molestations must also not be underestimated.

Way over 300.000 construction projects of different quality are available for architects and expert planners nowadays. During the planning process and subsequent calls for tender, relevant recommendations of ecological management must be considered also. Trades-people and construction companies must be contractually locked into clearly defined ecological criteria. The construction companies receive a clear guideline from the specific stipulations in the BOQs, regarding prohibited harmful substances and construction materials to be used.

For support of construction management and making matters easier on location control, an "Ecological manual for building management", for instance, is recommended. Such a brief guideline, small enough as a booklet to fit into a shirt pocket, summarizes all approved projects from a construction-ecological aspect as well as the most important trades.

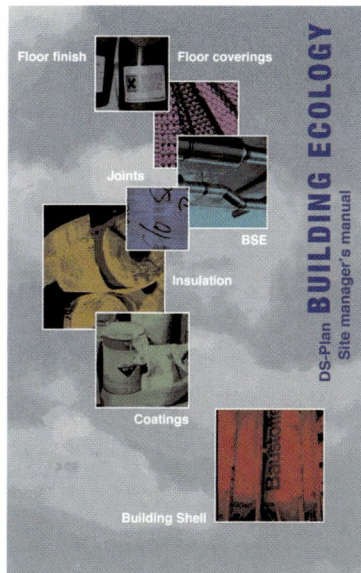

Fig. 3–77 Construction management manual
for construction ecology

3.6.4 Optimizing daily Operations

One elemental aspect of Green Building is to warrant the sustainability achieved during the construction and planning stages, through the use of intelligent control and automation systems, during running operation also. Small deviations from control technology of a building, already, are capable of reducing unnecessary costs and environmental stress over a longer period of time.

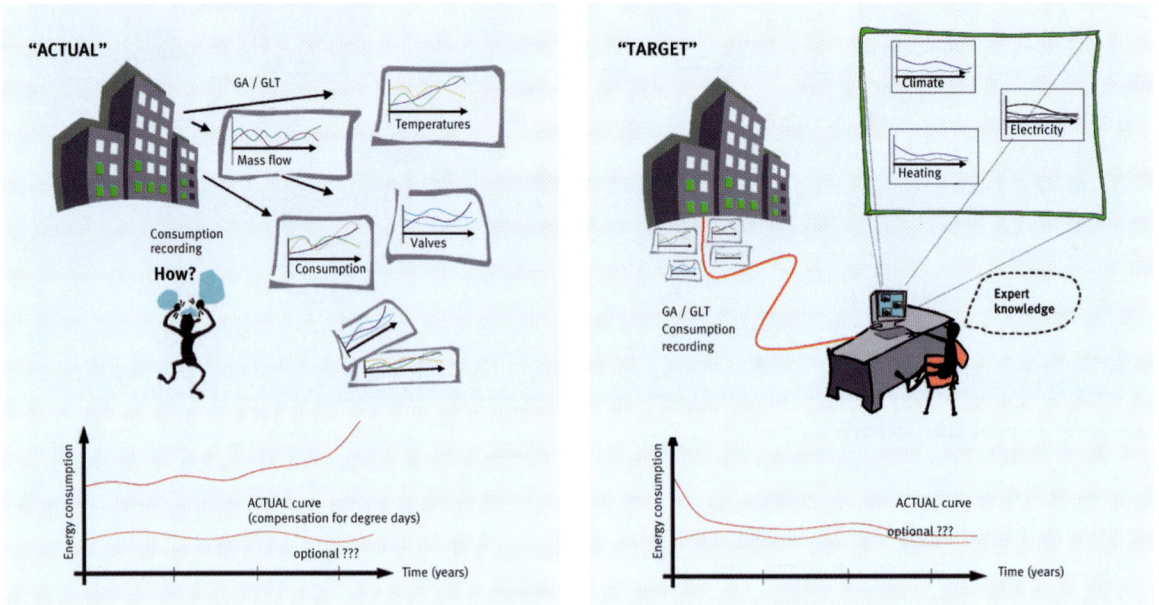

Fig. 3–78 Optimizing of operations today and tomorrow

This is why targeted control of indoor temperatures can lead to immense savings of energy consumption. Simulation and emulation allows for verification, in advance, of all aspects of building engineering and to optimize them.

Building foresight also includes a maintenance and cleaning plan. This way, running costs can be lowered and the life cycle of a building is increased.

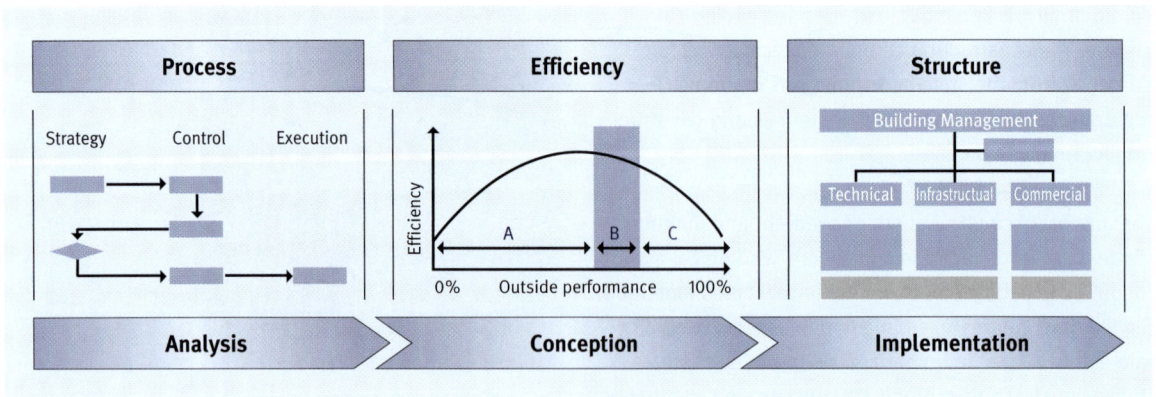

Fig. 3–79 Development of optimum administration organization

3.6.5 Going easy on Material Resources

The approach of sustainability acts on several levels during selection of construction materials: optimization of cumulated energy expenditure can either be undertaken through selection of construction materials with a high level of longevity or through use of construction materials made of renewable raw materials.

Use of regionally available raw materials is reflected in their low primary energy expenditure. Use of local woods for shuttering timber, for instance, additionally contributes to rainforest conservation. Ideally, the entire life cycle of a building, including renaturation, is already worked out during design of a building. This allows for selecting materials used under consideration of their subsequent further use already, and efficient recycling can be undertaken. Further, components that can be dismantled like windows, partition walls or ceiling panels and careful use of compound materials can later on avoid an expensive and material-intensive demolition process.

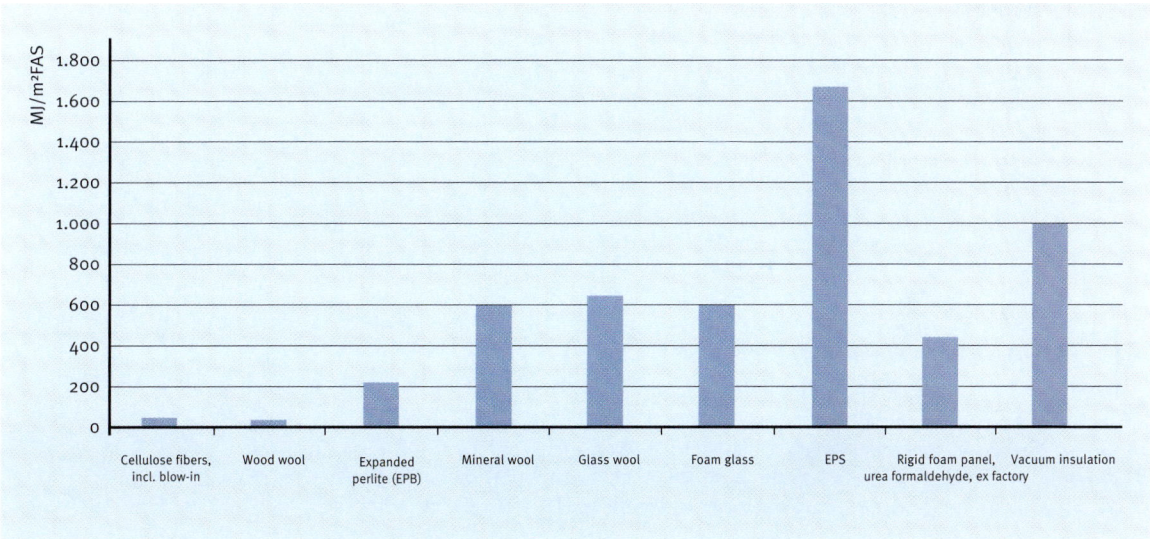

Fig. 3–80 Cumulated primary energy demand for alternative insulation materials

4 Tasks of Project Management

Regardless of whether the client handles project management autonomously or assigns it to external service providers – the more consequentially and extensive the various tasks are handled, the better the result is going to be. This applies in terms of function and architecture as well as economical optimization aspects, sticking to the specified cost frame and clear sequences with adherence to the agreed schedule. For this reason, we are initially going to look at all the required tasks without assigning these to a specific management strategy.

4.1   Demands on the Project Manager

Anyone wishing to become project manager for a large construction project needs to be a generalist. He/she must master the content information outlined in chapters one to three and also know about their interrelations and interfaces. This applies to process organization as well as target definition, the planning process and construction implementation. A project manager is not only responsible for processes to be undertaken as hassle-free as can be but he/she must also act as advisor to the client while also becoming a partner to the architect and even a coach of the specialist planners. For conventional projects, two significant errors are often made.

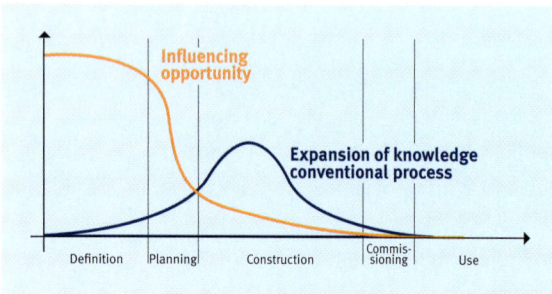

Fig. 4–1a Conventional process

Time is often wasted and the required knowledge expansion missed in the first phase of the project, the processing step called "target definition". One has the greatest possibilities just in this phase of influencing the project positively. The knowledge, however, is built up only in the planning phase, reaches its highlight during the execution phase and has largely disappeared again with commissioning.

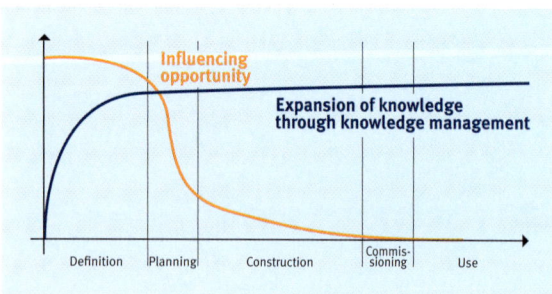

Fig. 4–1b Process with knowledge management

The situation behaves quite differently if an experienced professional and generalist is asked to handle the project management. A successful project manager will involve the project participants at the right (early) time, activate their knowledge and unite it to bring about the right decisions. The project manager must know the exact connection between the various process steps and have the know-how of the planners and executing companies and develop a solid knowledge management foundation from this. However, knowledge is only of use when it is applied purposefully and provided to all involved over a suitable project communication management.

In this respect, the hereinafter-described activities of project management only really take effect when "the boss" as a generalist is an accepted discussion partner who interacts on eye-to-eye level with the client and the architect. Such a professional will actively and personally steer the definition process, devise the first milestone plan, undertake the first rough-estimate cost calculations and initiate and accompany processes for the improvement of economical and quality considerations.

As managing director for a limited time, the project manager takes on leadership responsibility for several hundred employees. This includes own employees, those involved in planning, as well as staff in charge from of all of the executing companies. In case of a very large project, the total number of persons to be coordinated adds up to far over 1000.

Where does the secret of success lie? The project manager appears as a coordinator with integration capacity and motivates those involved to provide their cooperation from interests of their own. He/she creates the bases for an ordered organization sequence where those involved, as a rule, will approach problem solutions from self-initiative.

Activism dynamics are unsuitable as project managers since they tend to throw every project into chaos in a short time with their brash appearance in absence of any solid technical knowledge. But also introverted experts who shun confrontation and decisions are not made for the job.

4.2　Project Organization

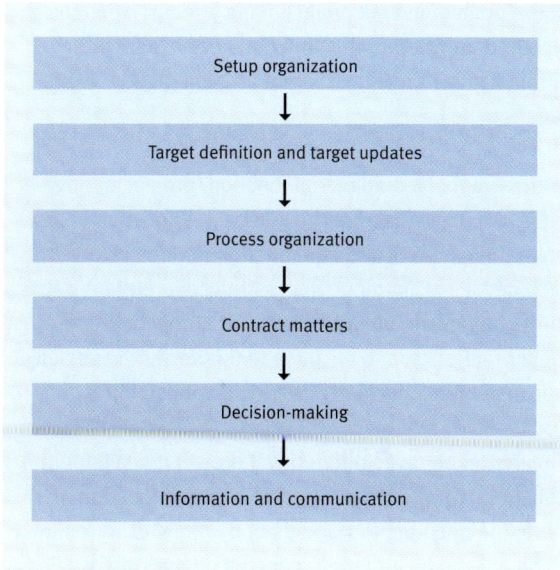

Fig. 4–2 Project organization

4.2.1　Client Organization

As a rule, all the common organization forms lie between a strictly hierarchical construction and the wanted or unintentional chaos.

The structure described as "group dynamics" distinguishes itself by the cooperation of "individualists" who are not organized in an ordered hierarchy. If a "leader" crystallizes,

the group members follow this person – or not. This is often dependent on whether his or her goals agree with theirs. Common undertakings are usually carried by enthusiasm and frequently lead to creative approaches but are left without result, as a rule. This type of organization is definitely suitable to work out project bases and define targets. However, it is useless for the actual project execution.

Experience shows that a clear and simple structure is necessary for an ordered execution of a project and that this needs to be independent of the preferred leadership method of the project manager. Every one involved must know whom to address if problems appear. He or she must also be clear for which areas and what staff members he takes on responsibility and to whom he can assign tasks.

These requirements are best met by the hierarchical structure. This is only valid, though, when this is built on task-related competencies and the corresponding autonomy is also delegated alongside the target-settings. In addition, the number of hierarchy levels must be kept as low as possible.

For all structures built up hierarchically, attention ought to always be paid that they do not lose their clear and simple makeup through ever more and new requirements and end up as "giant corporation" monuments. If a project manager already finds such unclear structures present at the start of the project, then he or she must try everything to "uncouple" the project.

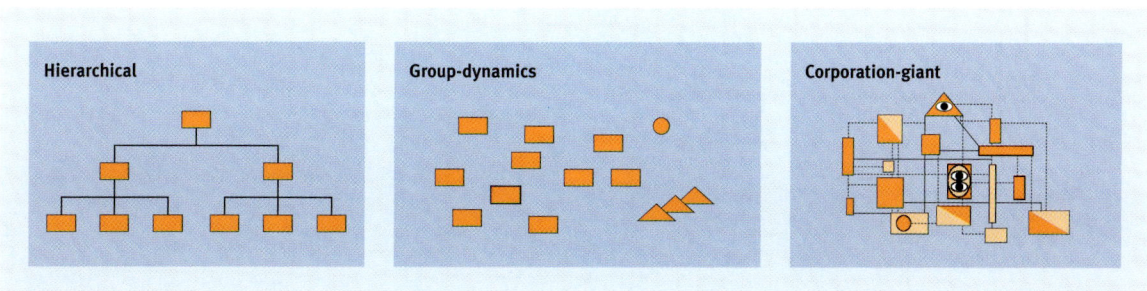

Fig. 4–3 Different types of organization

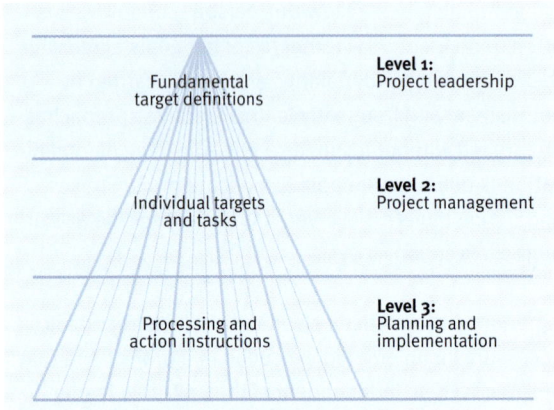

Fig. 4–4 Clear decision-making hierarchy

Project management is an indispensable instrument for the client, for the purposes of process coordination and cost control. Therefore, it should be established directly at the point of overall project management. It has to be weighed up whether this function needs to be arranged within a given project structure as a managerial or staff function, however.

In the event of a managerial function, project management can pass direct instructions and allocation of tasks on to the planning and construction entities involved, provided it is a matter of process and cost-relevant issues. In the framework of an efficient and stringent process, clear preference is to be given to this variety, especially because there are clearly defined responsibilities.

Whatever happens: A clear decision hierarchy is indispensable. It must always be ensured that the same structure is generally adhered to, from project leadership over project management up to process and action instructions. Only this approach warrants that the instructions clearly arrive at the executing person/s and that the process report is furnished also to project management within this structure.

The matter behaves differently if project management collaborates with the contractor in a staff function. In this case, the documents worked out by project control need to be passed on and then enforced by the employees of project management. Besides increased staff expenditure, this also results in a kind of indirect control with restricted autonomy. Project control works more as assistant of the various project managements and acts in an internal controlling function in this case.

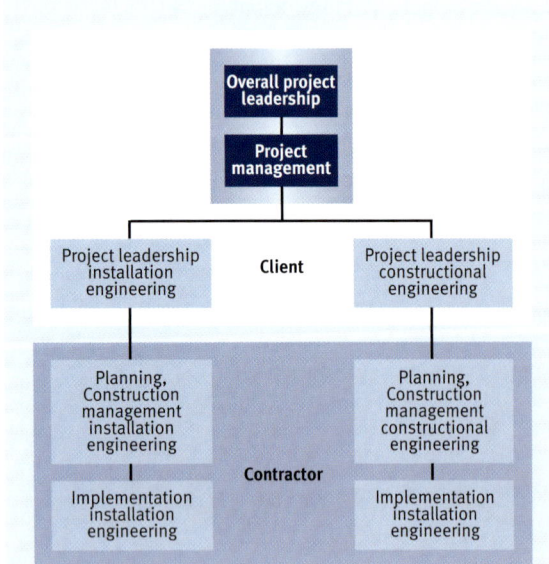

Fig. 4–5a Project management in managerial function

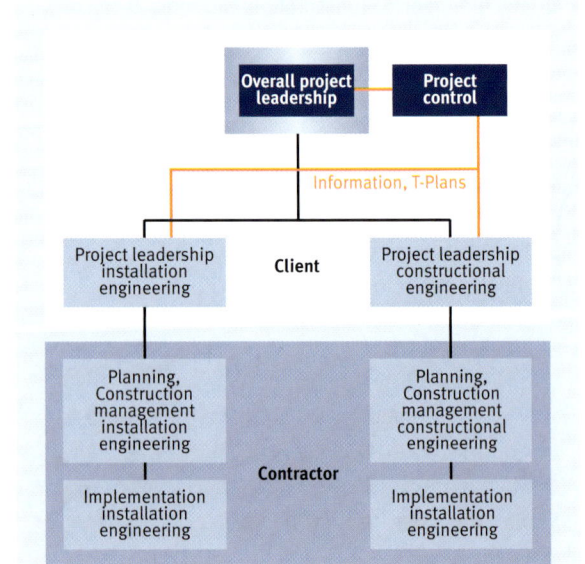

Fig. 4–5b Project control in staff function

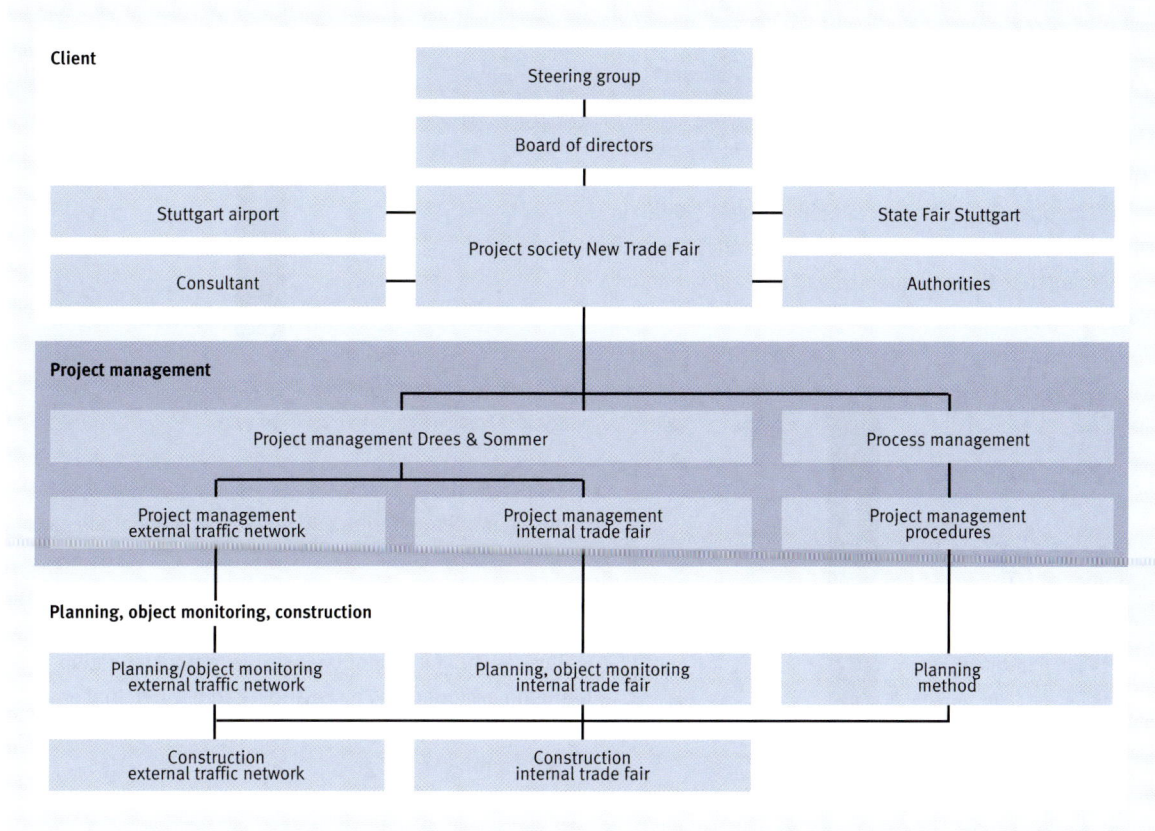

Fig. 4–6 Specific example for an organizational structure

For very large projects, the client organization is not always so simple to show. This applies primarily if, like in the case of the Fig. 4–6, a project company is set up for the construction of the Stuttgart state fair, which – so to speak – has several fathers and mothers. In that case, immense importance is placed on a clear decision hierarchy. In the case in hand, although the project company is supervised by a supervisory board and/or a steering committee, requirements and inputs originate from the users and operators whereas, for instance, the parking aspect is financed and controlled by the airport. The requirements of authorities and – primarily legal – advisers then also come in, especially when it comes to room arrangement and approval procedures.

The project management then must be organized, too, adapted to the laws of the project. So that everybody has the right contact person, the project management team is divided up into external traffic network, internal trade fair and, separate to this, method management.

Planning, object supervision and construction analogously to this are then also categorically structured.

4.2.2 Target Definitions and Target Updating

Clearly defined target-settings are the basis of every ordered, nevertheless creative, project execution. The basis of target definition is problem analysis. This means, initially, finding a solution to the question: "What do we want?" Frequently, it is simpler initially to ask: "What do we absolutely not want or only very reluctantly?" Unsatisfactory target-settings frequently lead to results that only insufficiently correspond to the initial task definition.

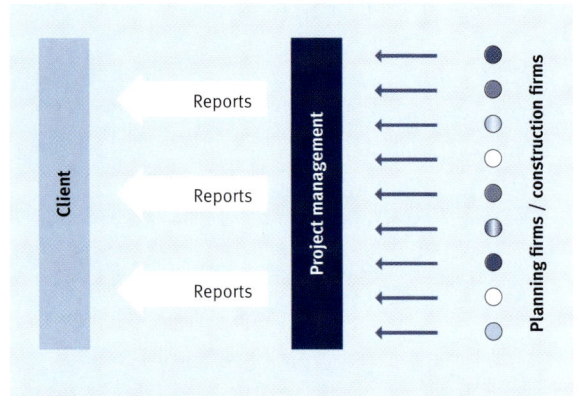

Fig. 4–8 Feedback on condensed information

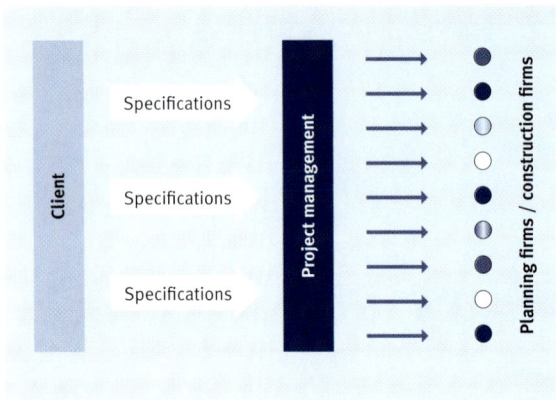

Fig. 4–7 Project management via target definition

The specifications of the project leader are passed on by project control to the various departments, planning leaders and site supervisors in the form of task definitions and schedules .In turn, these plan ahead and supervise the execution of planning and construction on the basis of the task definitions. A gradual improvement of the target definitions of the project manager arises from this procedure, across the individual hierarchy levels. This allows for the concentrated and essential target definitions of the project manager, which can be formulated without too much time being required for this undertaking, and are then implemented by a multitude of involved entities, in a clearly defined and scheduled manner.

The individual activities and their handling, just like any problems appearing in the process, are reported to project control by the planning and executive levels.

Project control then combines the individual reports into a concentrated report for the project manager, where

especially those problems are represented that appear during implementation and suggestions are made as to how to resolve them.

All these communication events are linear and multistage. The danger of information loss due to transformation cannot be excluded any more and it can happen that some target-settings are not taken care of according to the wishes of the project manager.

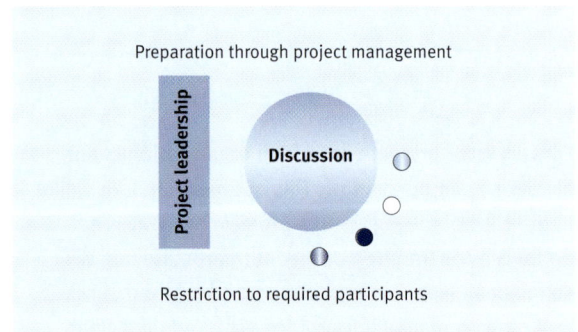

Fig. 4–9 Direct communication only for coordination need

It is therefore required, besides the linear system of the distribution of tasks via schedules or task catalogues with corresponding feedback, to also look for direct communication in the form of meetings with the project leader. Decisive is that these meetings confine themselves to the essential contents whose clarification requires the presence of the project manager. Each of these meetings must be carefully prepared, recorded and communicated by project control.

4.2.3 Project and Plan Segmentation

The organization of a large-scale project cannot simply be undertaken by expansion of conventional structures into one single monster organization. Rather, it requires segmenting into subprojects until easily comprehensible size orders and units result, which can be dealt with by the usual organizational methods.

One possibility is the division into a matrix organization with subprojects and subject or expert areas. In this, the subprojects are led respectively by a partial project leader, while the specialized divisional directors cooperate trans-sectionally with all the partial project leaders.

Of course, a superseding overall organization needs to be installed in addition (organizing the organization). This is also the reason that the organization of a large-scale project of > 100 mio. € requires considerably more effort than the organization of smaller projects. Through the additional organization level, specific effort required for very big projects rises exponentially and this one can only be handled through highest professionalism levels from all those involved.

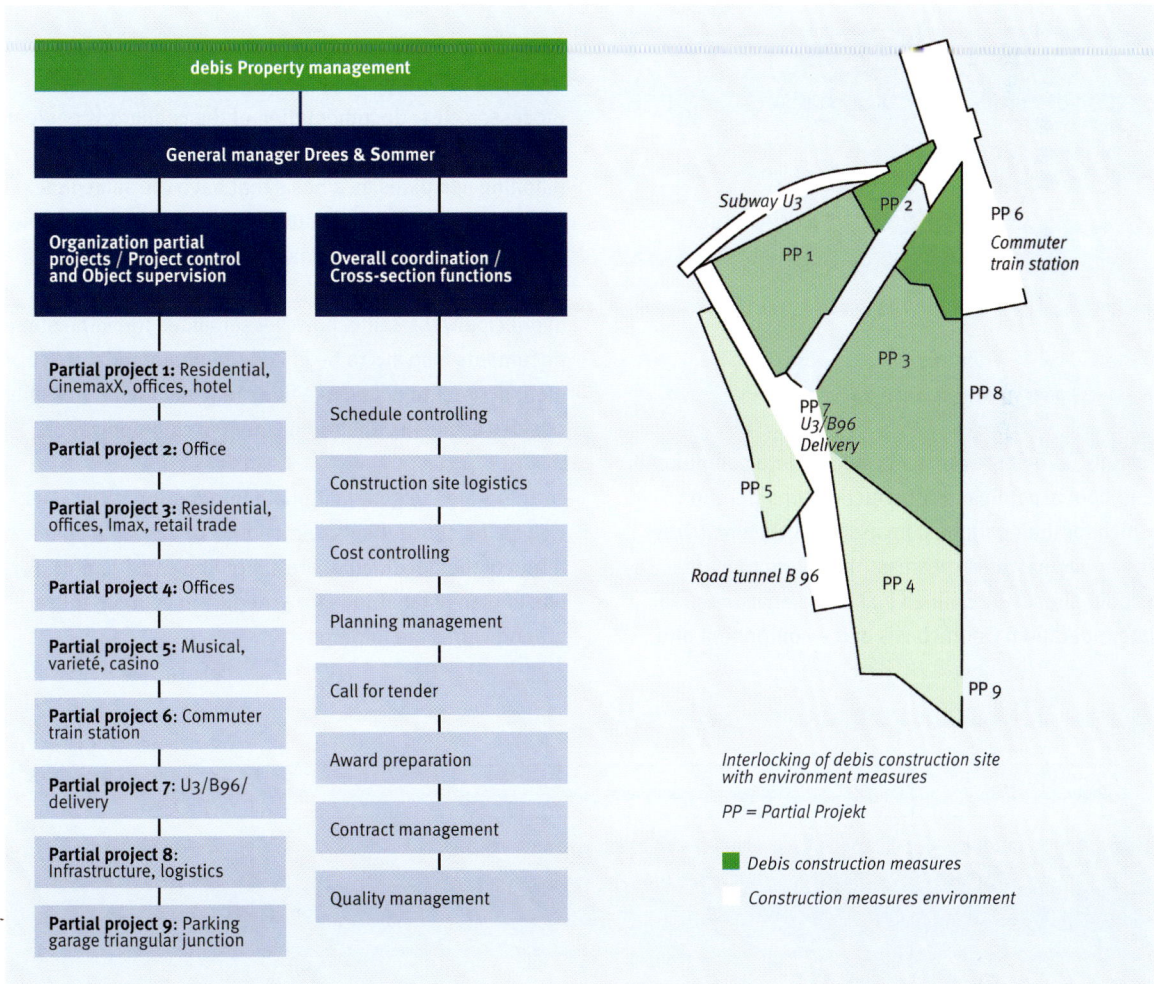

Fig. 4–10 Partial projects and cross-section functions

As the next task in the context of project organization, the basis for communication with each other must be established. This happens by the project being structured according to individual sections.

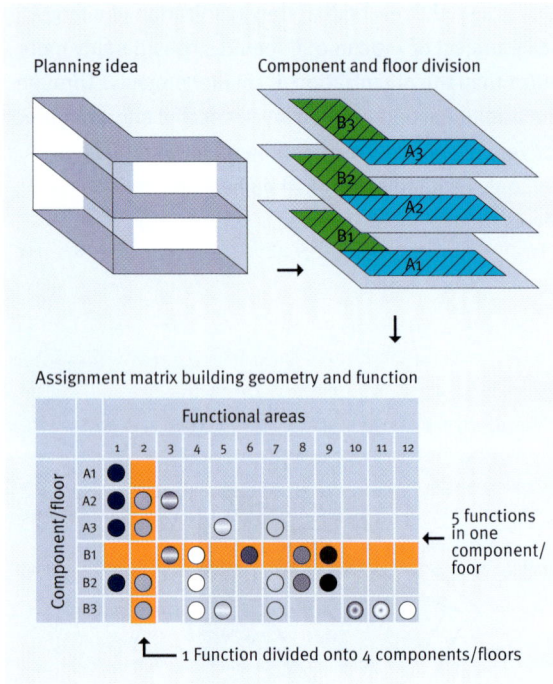

Fig. 4–11 Object segmentation and coding

For this, at first the project is split up into components and floor according to constructive consideration, which for their part must be assigned via a matrix to the individual function areas. These function areas should already be defined just like is envisioned for later operation. On the basis of the component and floor structure, a general plan division is carried out for all scales and all partial trades.

The information can further be clarified by a system drawing site plan. The respective part of the building represented is to be indicated either by a border or by background colors (A).

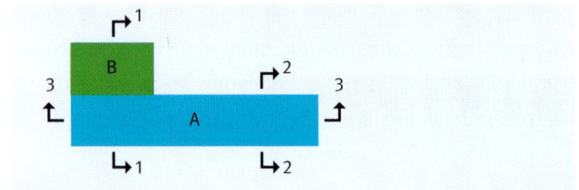

Fig. 4–12 System drawing site plan

Since all plan documents are – as experience shows – subject to change over the course of the planning process, a clear documentation of the changes is of great importance. It is important that all those involved in planning use plans as a basis that have the same index. An amendment documentation is built up approximately just like in the scheme update information service.

In principally the same, however simplified form, the other documents also are to be encoded (memoranda, letters etc.), in order to also make filing (e.g. microfilm) possible besides achieving an improvement in communication.

Furthermore the general update information service, as well as the subsequent maintenance by a clear documentation of the documents, are simplified considerably. For use of plan list methods for schedule execution, these ground-works are imperative.

Index	Date	KZ	Type of amendment	Initiator
01	22.01.1997	Oh	Gaps 9, 12, 15, 27 moved and/or enlarged	Engineer TA
02	17.02.1997	Mai	Wall in axis D/8-12 moved by 45 cm	Client/PM

Fig. 4–13 Scheme update information service

4.3 Contract and Risk Management

An essential prerequisite for an operating project sequence is a clear and, in regards to content, synchronized contract concept for planning and construction.

Precisely for large buildings with their multitude of different factors, it is imperative to know the possible consequences of contract changes for investment costs.

In the contracts, a thought-out system for the compensation of additional services or changed performances must therefore be included, besides the agreement concerning basic services to be provided. This avoids that single contractors can suddenly carry out almost arbitrary price corrections upward due to changes and a shortage of time. In the interest of a harmonious

cooperation, also, certain "rules of the game" should be agreed on for contract execution based on the principles of partnership and cooperation.

4.3.1 Planning Contracts

As the basis for contracts with planners and consultants, the HOAI = Honorarordnung für Architekten und Ingenieure (Fee Structure for Architects and Engineers = FSAE) is used. The basic services are listed in detail, inclusive of the individual fee percentages and fee panels which represent a mandatory fee entitlement via the HOAI. For partial basic services, after FLG, the Steinfort table can be used for orientation or the assessment tables after Simeon or Pott/Daahlhoff.

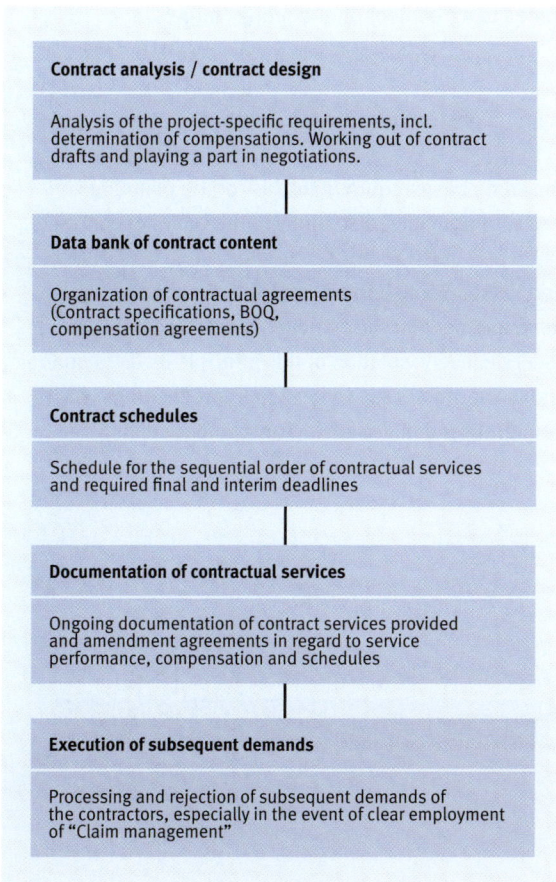

Contract analysis / contract design

Analysis of the project-specific requirements, incl. determination of compensations. Working out of contract drafts and playing a part in negotiations.

Data bank of contract content

Organization of contractual agreements (Contract specifications, BOQ, compensation agreements)

Contract schedules

Schedule for the sequential order of contractual services and required final and interim deadlines

Documentation of contractual services

Ongoing documentation of contract services provided and amendment agreements in regard to service performance, compensation and schedules

Execution of subsequent demands

Processing and rejection of subsequent demands of the contractors, especially in the event of clear employment of "Claim management"

Fig. 4–14 Sequence contract management

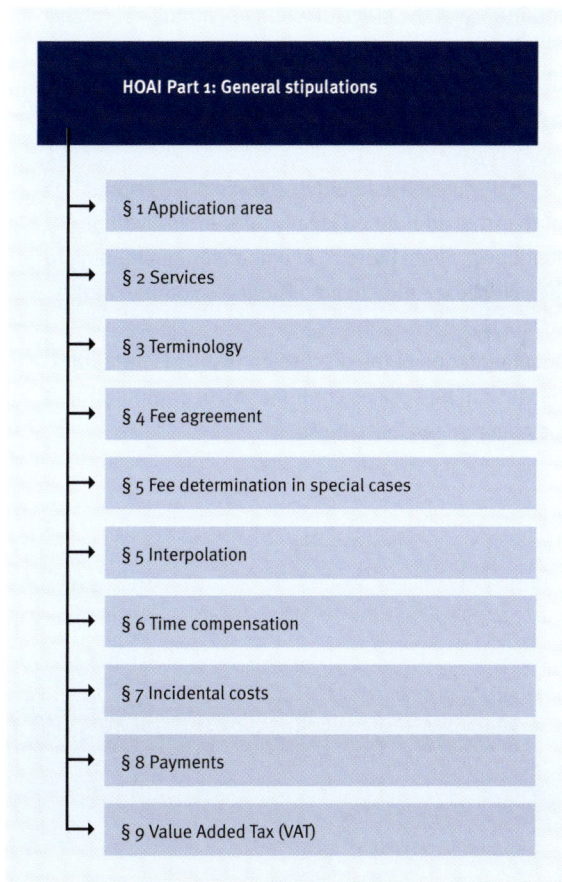

HOAI Part 1: General stipulations

→ § 1 Application area

→ § 2 Services

→ § 3 Terminology

→ § 4 Fee agreement

→ § 5 Fee determination in special cases

→ § 5 Interpolation

→ § 6 Time compensation

→ § 7 Incidental costs

→ § 8 Payments

→ § 9 Value Added Tax (VAT)

Fig. 4–15 HOAI segmentation part 1 – general stipulations

General terms and conditions for planning contracts:
Just as important as the correct performance outlines, we need to consider fair and adequate terms and conditions, which are not regulated in the HOAI. They should be uniformly created for all involved, whereby for different paragraphs a graduation should be carried out according to size and influence of the individual performances. These graduations will refer particularly to liability and guarantee but also to agreements on compensation. Here, also, one should take care not to create a gigantic work structure by lumping together as many terms and conditions as possible from various good and bad contracts, where the individual specifications contradict themselves constantly. In principle, the AGB law (Business Terms Law) must be taken into account for all contractual matters.

Corresponding schedule agreements are of special importance at the stage of writing planning contracts. The deadlines should not be set too tight, especially for the initial stages of the project, in order to be able to fully exploit the creativity of those involved.

Phase discussions should be agreed on at regular intervals, in addition to legally relevant individual deadlines, where there is a comparison of planning to the actual execution degree (Target-Is comparison).

Documentation of the discussion results is important since it can be seen as confirmation or continuation of the common business basis.

Altogether, we can confirm that the contract must be prepared as briefly as possible and in as much detail as necessary. For very large projects, bringing in a well-informed lawyer in this topic can be helpful. As a rule, the demand for a contract that is kept as short as possible is then only difficult to fulfill, though.

It is fundamental that clear contract conditions are established, which safeguard performance delivery and avoid later fee disputes . Such differences always lead to a reduction of performance. These clear conditions serve to improve cooperation between the client and the engaged planners and advisers quite considerably and avoid unnecessary losses from disputes and, if possible, these conditions should be in place immediately from the beginning of the rendering of services. In the event of excessive terms and conditions, an optimal performance delivery cannot be forced despite all care taken when it comes to agreement on the project targets.

4.3.2 Construction Service Contracts

The same basic requirements as for the planning contracts, in principle, also apply to the building contracts.

Here, also, clarity and unambiguity are the supreme commandment for the avoidance of unnecessary disputes. As a rule, every violation of this principle leads to quarrels and in the worst case to long court proceedings, which frequently end in unsatisfactory comparisons.

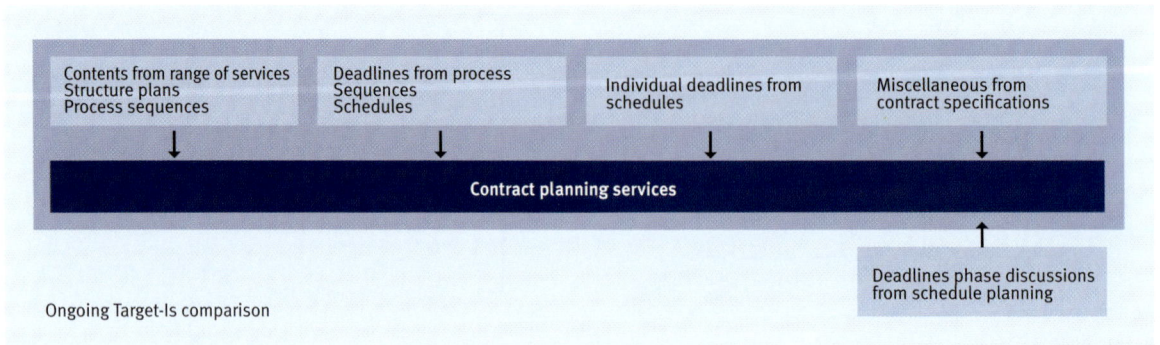

Fig. 4–16 Ongoing Target-Is comparison contents, appointments and deadlines

	Settlement contract (Unit fee contract)				Flat fee contract — Detail				Flat fee contract — Global / Simple			Flat fee contract — Global / Complex
	BOQ				**BOQ**				**BOQ**			**Construction outline (target-oriented)**
	Item	Number of architects	Service	Total price	Item	CON-amount	Service	Total price	CON-Amount	Service	Total price	
0% ▼	1	M_1	EP_1	GP_1	1	M_1	EP_1	GP_1	M_1			
	2	M_2	EP_2	GP_2	2	M_2	EP_2	GP_2	M_2			Usual range of services
100% ▼	n	M_n	EP_n	GP_n	n	M_n	EP_n	GP_n	– Recognizably incomplete – Completion clause for remaining services required			– Technically required – Functionality – Security
				Sum total				Sum total				
		Variable	Fixed	Variable		Variable	Fixed	Variable	Sum total fixed			Sum total fixed
Amount calculation:	Joint measurement				CON prior to execution				Rough estimate by CON + "Planning" for remaining services			Rough estimate by CON + "Planning" for total services
Risk:	PRIN				CON				Amount: CON Planning: Partially CON			Amount + Planning: CON

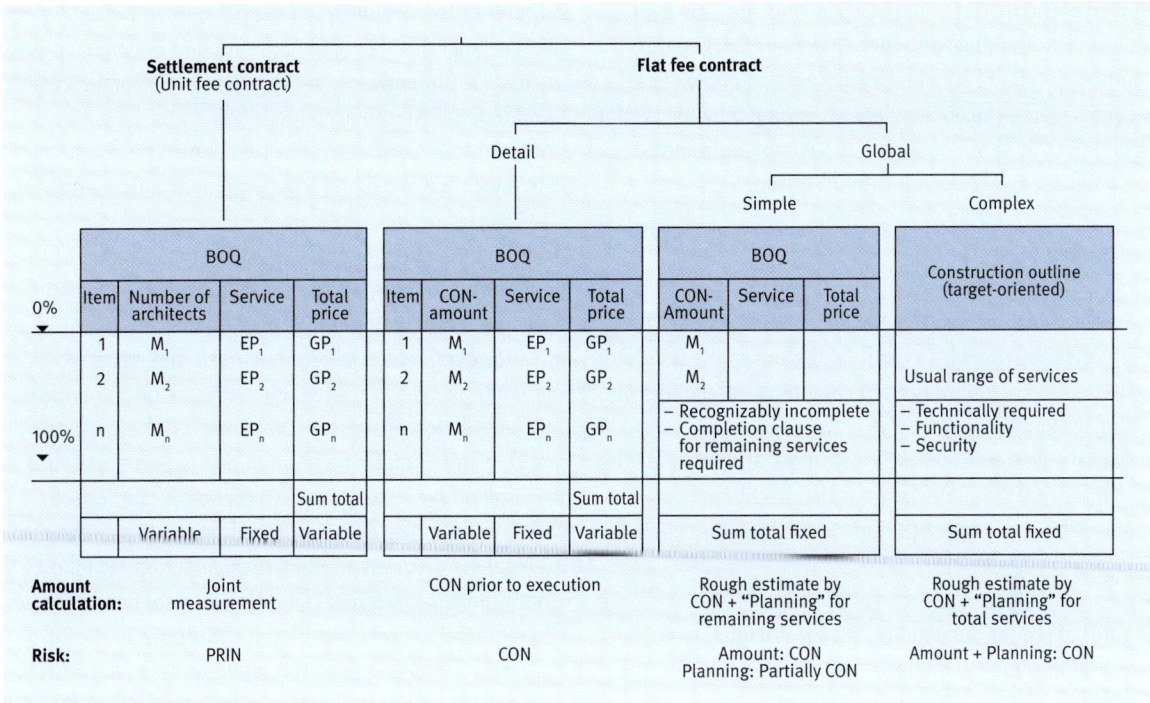

Fig. 4–17 Different versions of VOB contracts

Caution needs to be applied to unclear and therefore interpretable wordings in the tender documents as well as to doubtful flat fee agreements. This way, it is frequently attempted to pass on the risks from a not mature one-sided planning process to the contractor, particularly in the area of private clients.

Vice versa, the contractor in many cases makes use of every chance offering itself to assert subsequent demands (Claim management).

To prevent this, some customers have gone on to writing oversized terms and conditions into their contracts, which, in many cases:
– Are contradictory in themselves
– Massively either change or render ineffective in essential parts the VOB
– Due to regulations that are too detailed, end up requiring an entirely new set of regulations, which is then, however, neither recognized nor covered due to the complex connections

With consistent attention to a handful of principles, building contracts can be completed, offering sufficient legal security for both parties while not violating the principle of balance. Fundamentally, the VOB should be agreed on as a contract type for building contracts. Unlike BGB contracts, special arrangement rights of the client/CON are regulated here, which are absorbed by specific remuneration mechanisms as compensation (§ 2 VOB/B).

The illustration "Types of agreement for VOB contracts" represents the different types of agreement we must principally distinguish for VOB contracts. Depending on type of agreement, certain risks (quantity definition, planning) can be sensibly transferred to the construction company. Flat price contracts should be particularly considered when there is a likelihood of numerous changes in the event of accompanying planning. Global flat fee contract versions make sense if "prêt-a-porter" projects are to be constructed or if specialist company know-how is to be imported from the market.

Design of the BOQs: Some BOQ authors are only ever pleased with themselves when they have worked out complex solutions for simple problems. In many cases, we can observe similar behavior when preparing the tender documents. Without need, the BOQs are equipped with extra ballast and blown up with irrelevancies.

BOQs need to

- Be clearly structured (e.g. by lots)
- Be brief and precise in identifying the service required (If required by referring to the planning documents in the appendix)
- Contain clear rules for measurements and settlements, in the event that a deviation is planned from the specifications of the VOB part C for cases that can be justified

Control of tender documents: Tender documents must be verified by project management. For this, the approach illustrated in Fig. 4–18 is recommended.

This is independent of whether an award is to be assigned according to BOQ (individual award) or according to performance program (award to general contractors). Only by this type of control it is possible to guarantee that the targets set during planning are also realized accordingly, and that costly changes are not generated through the back door.

Adherence to VOB and VOL: In principle, the VOB (Award and Contract Ordinance for Construction Services) as well as the VOL (Procedures for the Award of Performances) should be taken into account. When handling public orders, this is mandatory. Private clients can forego the VOB A with the aim of the independent award negotiation.

According to its character, the VOB serves as an auto-generated economical law, not in the sense of a legal agreement but as General Terms and Conditions (AGB). It is, however, regarded as a balanced means between the building contract parties. It serves as decision base for construction processes, including its extensive commentary and jurisdiction granted by the Supreme Court. If a legal problem results, you are already confronted with it prior to the trials or while either justifying the base for legal action or responding to such action, as long as a comprehensive new legal situation has not been created by an individual agreement. Since this only applies in the rarest of cases, the VOB B, at the very least, ought to be declared an absolute contractual basis; the VOB C is therefore agreed on automatically.

There are many cases where additional General Terms and Conditions are used to try to bring about a one-sided amendment to the specifications of the VOB in favor of one party to the contract. Such interventions in the nuclear content of the VOB through supplementary contract clauses (written form clauses, formal com-

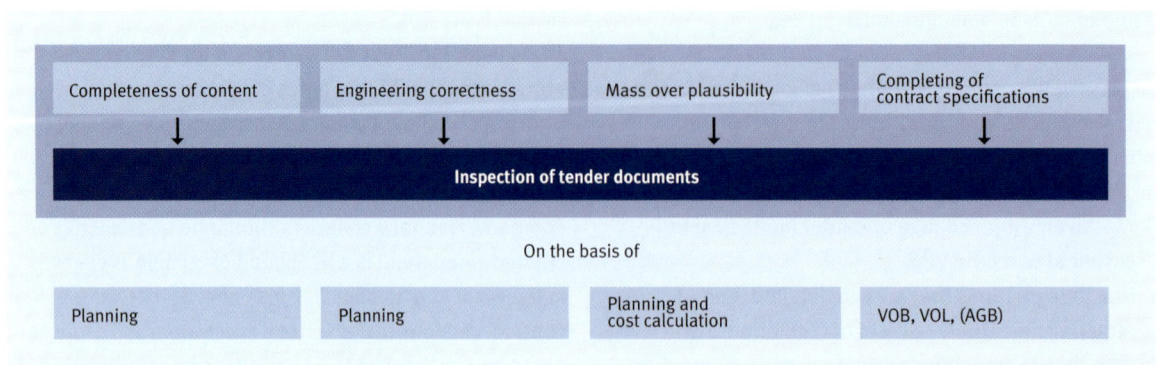

Completeness of content	Engineering correctness	Mass over plausibility	Completing of contract specifications
↓	↓	↓	↓
Inspection of tender documents			

On the basis of

Planning	Planning	Planning and cost calculation	VOB, VOL, (AGB)

Fig. 4–18 Inspection of construction service contracts

pulsions etc.) undermine the VOB and there is the danger that, via the regulations of the AGB law, the VOB is altogether dropped as a base for business.

As a contractual basis, merely §§ 631 f. of the BGB is available, which generally places the client in a worse position than the stipulations of the VOB.

4.3.3 Preventative Risk Management

Construction projects are subject to a number of influential factors that can gravely impact the economic considerations of a project.

Conventional project management orientates itself on indicators, primarily. In contrast, latent project target deviations can be included in the consideration, from project beginning onwards, by an integral risk management .

This proactive risk strategy is particularly in the Anglo-American world already standard by inclusion of the so-called Risk Costs. The relevant methodology is represented by the work steps indicated in Fig. 4–20:

The individual risks are identified in special project workshops and then put together in so-called risk registers that should not display more than 150 pieces. Evaluation is quantitatively undertaken by expert assessment, where size can be sensibly reduced to the essential management

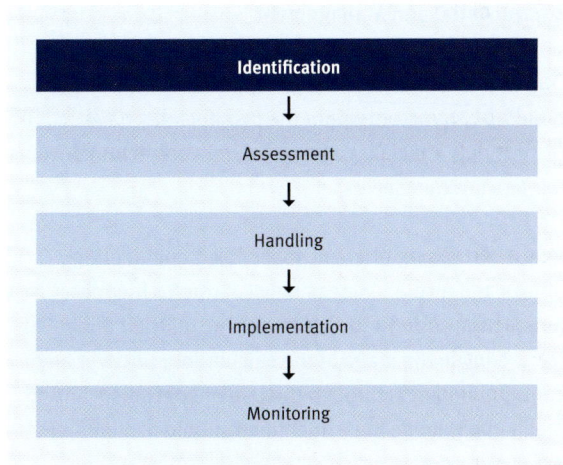

Fig. 4–20 Proactive risk-strategy

focal points through prioritizing (e.g. ABC analysis). Risk mastering through quantitative chance management is the focus here, to show suitable alternatives of action for those in charge of the project managers and to be able to jointly implement suitable strategies.

Accompanying the project with regular risk controlling (base: data base) makes sense, combined with suitable reporting for the strategic leadership of the projects. Particularly, gradual risk reduction due to chance management has to be documented, to show the success of the measures decided.

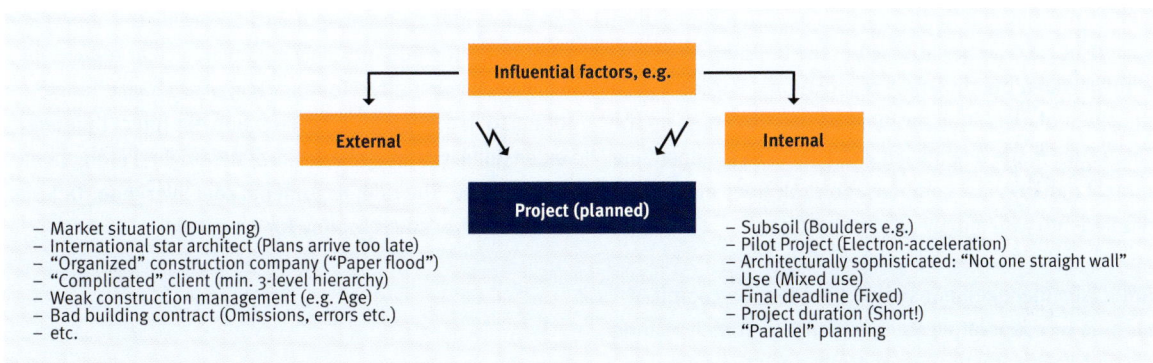

- Market situation (Dumping)
- International star architect (Plans arrive too late)
- "Organized" construction company ("Paper flood")
- "Complicated" client (min. 3-level hierarchy)
- Weak construction management (e.g. Age)
- Bad building contract (Omissions, errors etc.)
- etc.

- Subsoil (Boulders e.g.)
- Pilot Project (Electron-acceleration)
- Architecturally sophisticated: "Not one straight wall"
- Use (Mixed use)
- Final deadline (Fixed)
- Project duration (Short!)
- "Parallel" planning

Fig. 4–19 Risk factors for construction projects

4.3.4 Anti Claim Management

Due to the increasing complexity of building contracts, building contractors within the last decades have increasingly specialized in exploiting contractual uncertainties for themselves .

Intensive training offers of specialized engineering offices and also an increasingly contractor friendly jurisdiction support the building firms in their endeavors to, already in the stage of the calculation, uncover any contract problems through the use of specialists and to preliminarily judge them. Identified supplement potentials are exploited to put specific dumping prices on the market (sub-value offers). Through claim management, achieving the target annual figures is striven for during construction via enforcing numerous supplements.

The mass profit generated from contract changes forces many clients into a specific anti claim management under increasing cost-pressure. Through formal weakening of the demands in legal and methodical weak points of the claims, construction quarrels swiftly result, which – especially for clients with deadline problems – give way to quick concession to uneconomical comparisons.

A specific anti claim management therefore represents only a short-term success optimization since construction firms know about the time consequences of their claimed demands. This concerns primarily hindrances which can be assigned to the risk area of the client and cause considerable construction time extensions and which do not seldom have 20 to 50 % of the original contract sum as building contractor target value. But also technical supplements are regularly part of the building time delay with so-called building time reservations. Particularly in the event of deadline-critical projects, this leads, on the client side, to the dilemma that, at the point of impending project timeout, fast compensation solutions must be entered into with mostly adverse results.

Anti claim management should therefore be complemented by cooperative management. In a timely manner, accompanying the project, contract deviations are causally documented, processed transparently and

talked about comprehensibly in regular cooperation conversations. Legitimate claims should immediately be paid. Controversial facts are negotiated on the basis of construction facts with a neutral authority in constructionally acceptable risk areas and, in the case of emotional disputes, are settled in a timely manner by suitable methods of conflict management.

The intelligent investment into the avoiding of construction-related disputes, with project management according to partner-based implementation philosophy, usually results in significant reduction of claim volume.

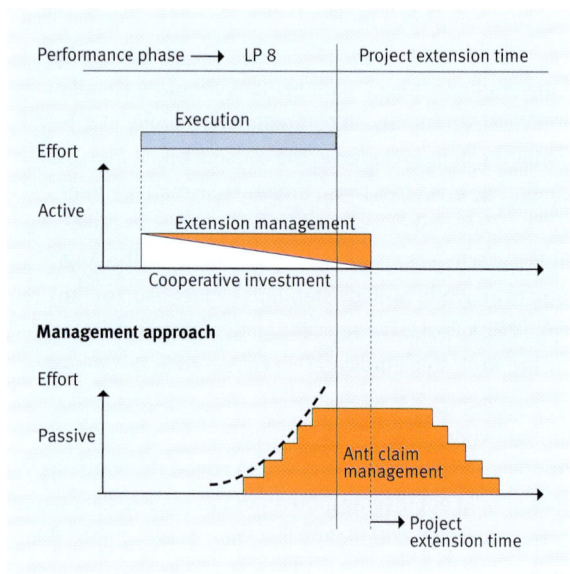

Fig. 4–21 Possible reduction of claim volume

An active management approach according to cooperation criteria, in the end, not only results in a clear reduction of management effort for handling of contract changes but also in a considerable reduction of project timeline extensions with the possibility of the original project time compliance. The rule: **Arguments cost everyone. Talking to each other saves money for everyone!**

4.4   Schedule Management

4.4.1   Network Planning as Simulation of Planned Procedure

In many areas, the network planning technique is employed as a planning instrument for time sequences. While, for simple and linear processes, a simpler time planning (Gant chart) usually suffices,also, project management cannot do without this aid at all due to the complicated contexts of the overall project and the large number of events and people involved .

The network planning method represents the only opportunity to clearly describe the connections of the various structures and individual events and to simulate and optimize the process for different alternatives. While the methods of computation are taken as being known, we shall here more deeply delve into systematics in the context of application for project management.

It is a simulation model that, once erected, makes an optimization of the processes and their permanent supervision possible by means of different parameters. The individual steps are explained on the following pages:

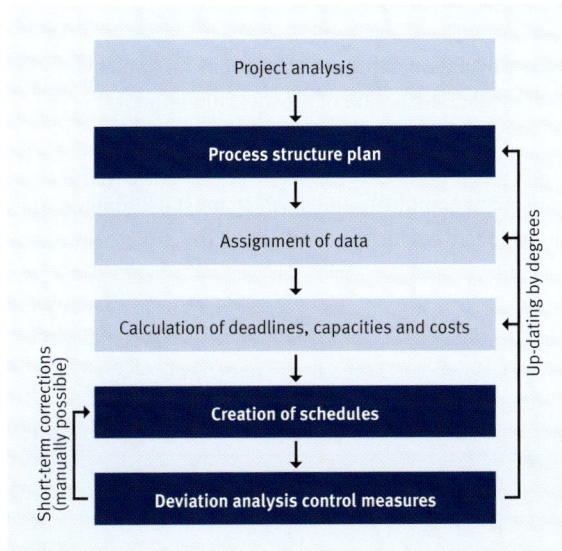

Fig. 4–22 Network planning method = simulation of schedule process

Project analysis: The first step prior to entering into the simulation model is a thorough analysis of the people involved, their tasks and ideas as well as the boundary conditions given. This analysis of the necessary events requires us, in principle, to ask the following questions:

– Who needs to fulfill what task?
– What precisely needs to be done?
– By when does it need to be done (the latest)?
– Where do the checks need to be done?
– How is handling to be undertaken and what do the checks look like?
– Why have target definitions of this project been directed precisely at where they are? Is there an alternative approach?

The ideas and demands of the client must particularly thoroughly be analyzed, for instance. Bridge financing of buildings, beginning with the acquisition of the property and concluding with commissioning, requires very high charges. The client will therefore always endeavor to keep construction time as short as possible in the context of the means available to him. These costs of the intermediate financing can be decreased by, for instance, single stages of construction being commissioned early.

Aside from the type of object the conditions of the construction site and the manner are also extremely important to the construction as well as manner and size of the development. The following points are examined among others, without claiming completeness:

– Position of Building (e.g. traffic conditions, construction site access, areas available for building site setup and construction material storage)
– Overbuilt areas, number of floors and m³ used renovated area for rising structures
– Type of construction, especially uniformity of component and degree of difficulty for formwork
– Division of the construction object into building parts via joints
– Type and degree of difficulty of foundation
– Options of connection to public supply facilities (e.g. district heating)

– Position of central air conditioners and, if applicable, recooling plants
– Demand for dust freedom for installation of electrical systems
– Special acoustics requirements

Only when this "Information gathering" has been concluded at sufficient extent, structuring can commence.

Structuring of the processes: The project manager must take into account adequately the interests of all involved in scheduling, in order to ensure the correct structuring of the processes. Disregard of this basic principle results in little readiness for constructive cooperation of those involved.

Process Planning must be an instrument that:

– Minimizes friction during planning and execution and
– Renders planning and implementation processes predictable

Therefore, realistic ideas and not wishful thinking need to be in the foreground of the construction of process structure.

The structuring of the processes requires greatest experience from the project manager, he/she really must know the interactions of the various processes and services in order to create a corresponding outline and accurately include the sequence of these events in his plans. Pre-structured standard processes offer assistance in this, for certain sections which represent a kind of operator's manual when provided with corresponding descriptions. For instance, for certain types of implementation, particular trade sequences are required or at least sensible, independent of the type of construction undertaking.

Calculation of time deadlines for milestone plans:
Relatively rough approaches to time planning suffice for the preparation of primary schedules. For example, shell construction work can be assessed according to time values as are shown in the following graph.

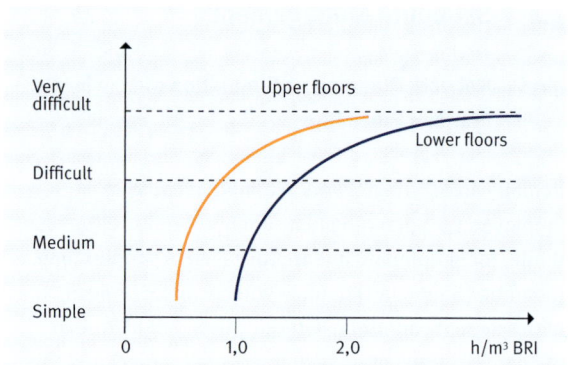

Fig. 4–23 Rough time values shell construction

Calculation of time specifications for the planning phase:
While, for the structure considerations for process planning, rough reference values still suffice for time and capacity considerations, a more exact calculation must be carried out for the continuation of the structure plans at medium and short-term levels to in order to obtain useful control schedule plans. The calculation of process timeframes for planning is carried out under consideration of project-specific boundary conditions like:

– Repetition level
– Degree of difficulty
– Routine formation
– Applicability of technical aids (e.g. CAD systems)

Time calculation of time specifications for implementation processes: Calculation of process duration for the individual trades of construction is carried out on the basis of time values per performance unit. The time values are empirical values, which must be adapted to the respective special conditions of the project.

Service	h/E	h/1000 €	h/m² €
Plaster works Wall plaster MP 75 Ceiling plaster	0,20 0,25	6 8	0,30

Fig. 4–24 Time values extension works

Calculation of process durations, when looked at in isolation, does not suffice for a simulation as realistic as possible of the later processes. The possible logistics system must rather also be taken into account when specifying timing. Among other things, this includes:

– Transport options to the building site
– Manner and scope of available transport means for vertical transport (especially during the development stage)
– Traffic routes in and around the building for horizontal transports
– Storage areas inside and outside the building

In the area of logistics, particularly in downtown settings, cost and time reserves for buildings are considerable and generally used too little and/or, in the event of non-adherence, there are high risks as far as ordered process execution is concerned.

4.4.2 Calculation of Deadlines, Capacities and Outflow of Funds

For network plan technique, individual performance is defined in accordance with the process structure plans as events, connected via arrangement relations.

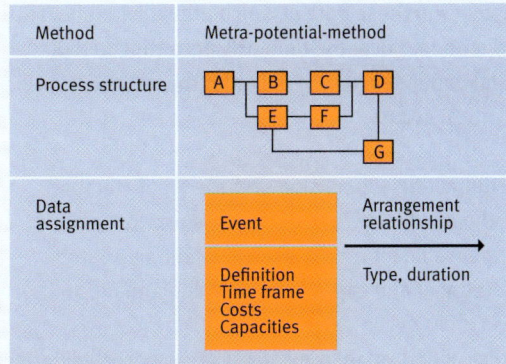

Fig. 4–25 Metra-potential-method

Each event is assigned a certain set of data like, for instance:

– Event duration
– Event costs
– Resources
– Organization-codes

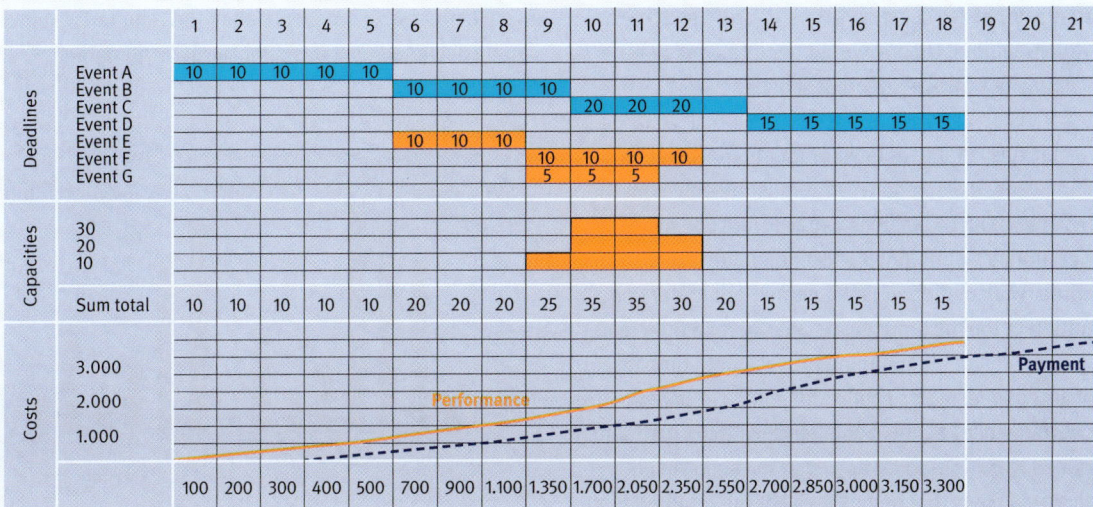

Fig. 4–26 Deadlines, capacities, costs

The scheduled process structure is calculated with the assistance of these data in which the individual events are distributed over the time axis according to their mutual dependencies. Defining for the sequence are the critical events.

For control and supervision at the level of medium and short-term time scheduling, the network plan is indispensable for the application of differentiated appointment and capacity resources planning.

Process optimization by means of capacity considerations:
The sequence of construction projects shows, again and again, that the provision of sufficient capacities for planning and construction is decisive for compliance with the schedules.

In many cases, and particularly for projects with very tight deadlines, construction progress is determined by the provision of implementation plans. A close co-operation with the firms that prepare these plans and a good coordination of the firms with each other are the prerequisite for compliance with tight deadlines. It is necessary to coordinate plan delivery and hence, construction sequence with available and/or sensibly applicable capacities.

The figure titled Schedule deviations through capacity shortfalls, for instance, shows that, in order to achieve a certain deadline, t2, one would need to employ at peak capacity a total of 28 processors.

The capacity available at most amounts, however, is 18 employees, a discrepancy that must be solved in the process plan. Either, planning capacity must be increased at short notice by bringing in additional firms or else a schedule recalculation must be carried out under consideration of the maximum capacities, which then leads to an extension of the planning process until time t2. It is not acceptable, in this context, to reduce required planning performances per se. Compliance with the predefined deadlines which would be achieved through this, is obtained via plans that do not yet have the required planning maturity (preliminary deductions).

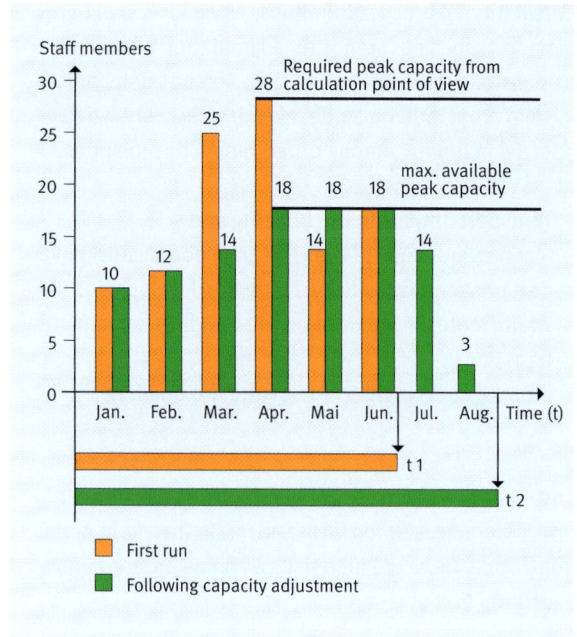

Fig. 4–27 Schedule deviations through capacity shortfalls

In consequence, these appear:
– Quality decline
– Additional costs
– Loss of time

Capacity examinations during the execution phase:
To each process of a network plan, the required resources can be assigned as a basis for capacity planning.

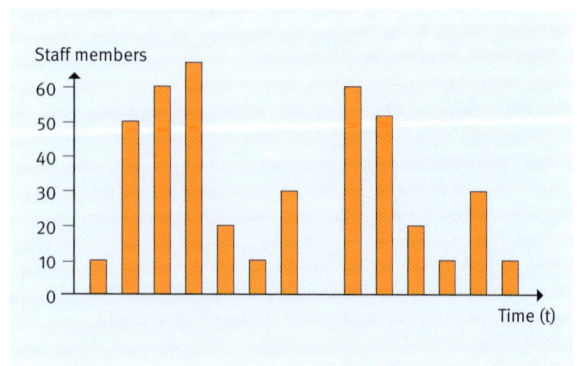

Fig. 4–28a Peak capacities in the event of early start

A summation of these resources in the event of early start or late start consideration normally yields a very arbitrary and unbalanced capacity line.

Since occupation of staff as continuous as possible or full exploitation of machines is, however, striven for as a rule, one needs to undertake capacity balancing.

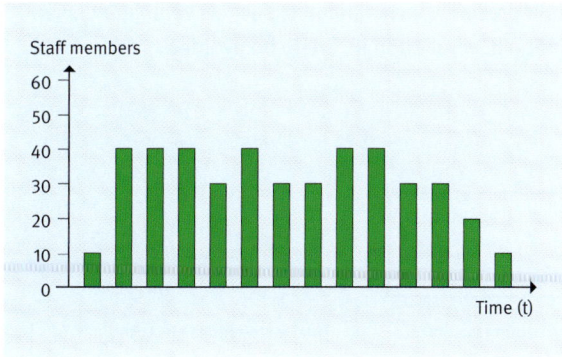

Fig. 4–28b Capacity balancing though buffer exploitation

As a first step, one tries to arrive at capacity balancing in the area of the calculated buffer times by moving events that do not lie on the critical path. When required, the critical path is extended in a manner to allow capacity limits to be adhered to.

Control of the outflow of funds (payment plans):

Interim financing costs (normally approx. 6 to 8 % of the manufacturing costs) constitute a considerable share in the total costs of a given construction undertaking.It is the task of the project management ,via corresponding simulation models, to achieve an optimum of construction time, building costs and expenditure development. Building on time scheduling, the network plan can also be used for determination of the expenditure course. The accompanying costs are assigned to the individual process, in which the cost amount is assumed over the duration of the process as being linear.

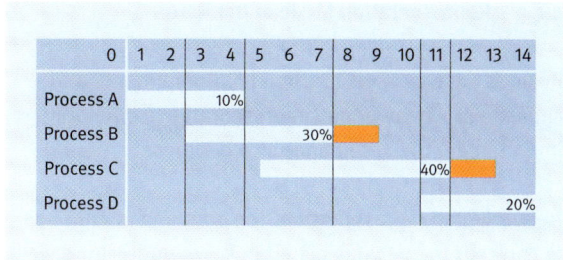

Fig. 4–29a Schedule for early and late start

The costs of the individual processes are summed up, whereby the sum curves for early start and late start of the processes distinguish themselves.

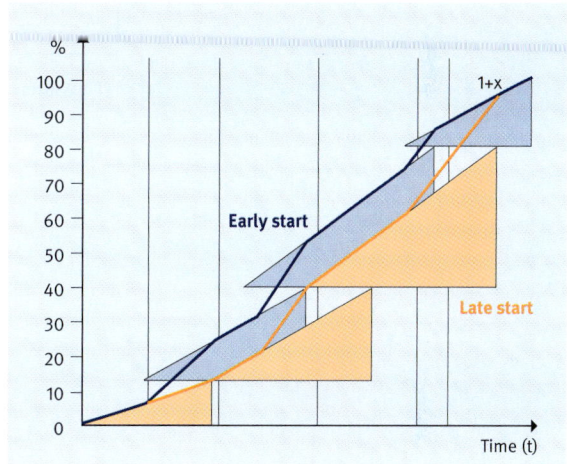

Fig.. 4–29b Expenditure sequence for early and late start

One can specify the probable cost course by using these curves and make them the basis of means management.

Another possibility for the narrow adaptation of cash means to be provided, versus actual expenditure course, is payment according to performance and appointment-dependent payment plans for the individual trades at a set payment day per month. The payment plans are pre-pared on the basis of the network plans and supervised by the project manager. The possibility of saving interim financing costs through skillful organization of the process has already been pointed out.

4.4.3 Representation Mode of the Plans

Besides the exact specification of appointments and capacities, a clear graphic representation is primarily important for the acceptance of schedules.

Network plan representation: The planning instrument of choice is the network plan that is not understood well and therefore not accepted many by parties involved in construction, though. This representation form, hence, usually remains the planning means of the project manager unless the network plan is structured, clear and easy to understand. Fig. 4–30, for instance, represents an excerpt from such an easily understandable network plan, as can be created for sections. It shows primarily the interdependncies of the processes. It is important that the events are described understandably so that one immediately and obviously knows what it is meant to represent. Unclear names, such as ventilation I, ventilation II, ventilation III, are to be avoided.

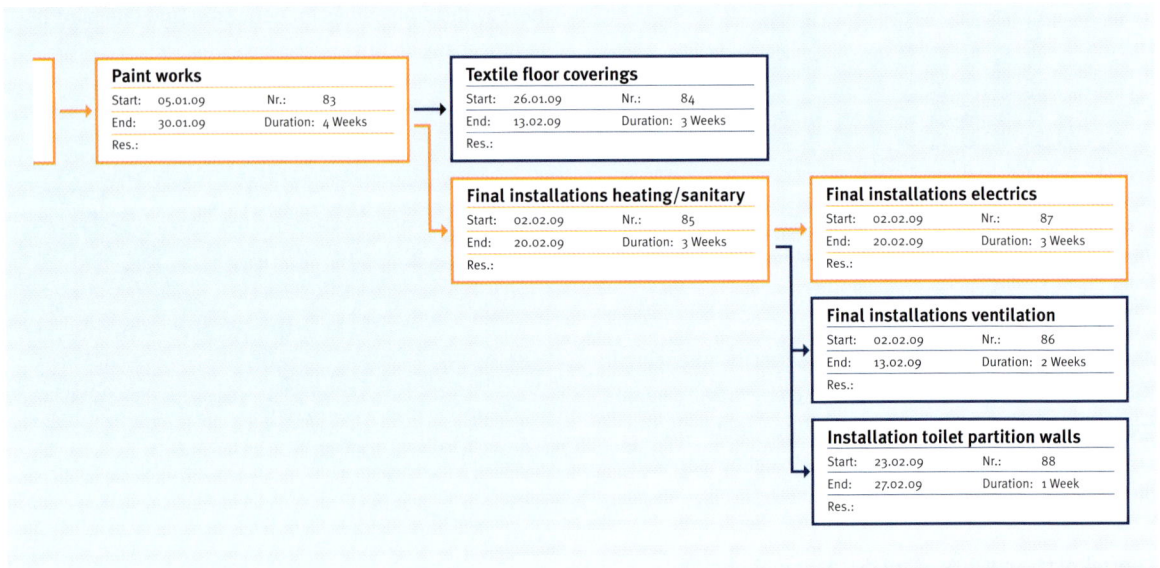

Paint works			
Start:	05.01.09	Nr.:	83
End:	30.01.09	Duration:	4 Weeks
Res.:			

Textile floor coverings			
Start:	26.01.09	Nr.:	84
End:	13.02.09	Duration:	3 Weeks
Res.:			

Final installations heating/sanitary			
Start:	02.02.09	Nr.:	85
End:	20.02.09	Duration:	3 Weeks
Res.:			

Final installations electrics			
Start:	02.02.09	Nr.:	87
End:	20.02.09	Duration:	3 Weeks
Res.:			

Final installations ventilation			
Start:	02.02.09	Nr.:	86
End:	13.02.09	Duration:	2 Weeks
Res.:			

Installation toilet partition walls			
Start:	23.02.09	Nr.:	88
End:	27.02.09	Duration:	1 Week
Res.:			

Fig. 4–30 Excerpt network plan

Representation of the process in a gant chart: Gant charts are still the most clear manner of presentation for construction processes. In gant charts, also, the works to be executed are displayed in a temporal order and duration of the individual work is specified on the time axis by the length of the accompanying bar.

Gant charts have the great advantage of immediate vividness. Therefore, all network plans should be represented, after calculation, as gant charts.

Gant charts, however, are not only a great display tool but also an excellent planning procedure. The specialist engineer who will be assigned project control is going to use them especially when it comes to specifying framework deadlines, provided he has sufficient experience to recognize dependencies.

Also, they render good services when it comes to cyclic processes for the repetition of work sections. If gant charts are used for the preparation of process plans, then one is also forced to fix deadlines when rendering the individual work processes, something that, incidentally, is also of advantage. The project controller immediately recognizes the time, when developing the project sequence, at which a certain measure must be implemented. This allows him/her to take into account unusual features that have an effect on sequence e.g. Christmas period, vacation season and bad weather days.

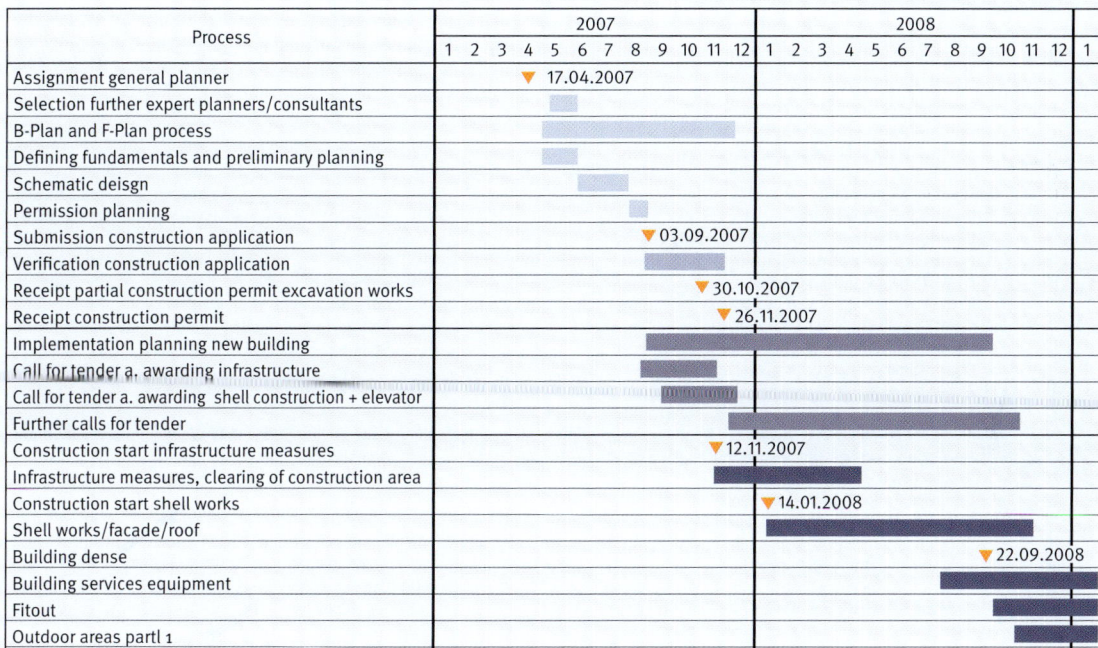

Process	2007												2008												
	1	2	3	4	5	6	7	8	9	10	11	12	1	2	3	4	5	6	7	8	9	10	11	12	1
Assignment general planner				▼ 17.04.2007																					
Selection further expert planners/consultants																									
B-Plan and F-Plan process																									
Defining fundamentals and preliminary planning																									
Schematic deisgn																									
Permission planning																									
Submission construction application								▼ 03.09.2007																	
Verification construction application																									
Receipt partial construction permit excavation works									▼ 30.10.2007																
Receipt construction permit										▼ 26.11.2007															
Implementation planning new building																									
Call for tender a. awarding infrastructure																									
Call for tender a. awarding shell construction + elevator																									
Further calls for tender																									
Construction start infrastructure measures									▼ 12.11.2007																
Infrastructure measures, clearing of construction area																									
Construction start shell works													▼ 14.01.2008												
Shell works/facade/roof																									
Building dense																					▼ 22.09.2008				
Building services equipment																									
Fitout																									
Outdoor areas partl 1																									

Fig. 4–31 Excerpt gant chart

Representation of the deadlines in a schedule list:
Schedule lists, sometimes also known as time tables, have primarily proven themselves because of their good readability especially for the specification of interim deadlines appointments and closing dates in tender documents. They are applied in project control for the specification of delivery dates for clearly defined services e.g. like plan deliveries. They are also suitable for the specification of completion deadlines for individual parts of a building and stages of construction. Ground plan schedules where, for instance, assembly appointments are entered for fitout works, represent a special form of these appointment lists. Such plans can easily be read and understood by every fitter. Appointment lists are therefore less suitable as a planning method; they represent, rather, an excellent method to make oneself understood to those involved in construction.

The possibilities for list printouts are many. From individual, unsorted lists up to categorized itemizations according to certain guidelines, you can choose as you please. Here are some of the most common printouts:

– Sorted by sequence of ascending buffer times
– Sorted by earliest possible starting date
– Sorted by special features (e.g. by trades)

For sorting according to special features, key numbers (codes) must be created for the processes, which are assigned to these features. One should consider whether ranking of the individual codes is possible so that e.g. one can indicate, as a superseding sorting concept, the components of a project and then carry out sorting according to segments within the individual, specified components.

Sample project

Nr.	Process name	Responsible	Duration	Start
1	**Definition of fundamentals / preliminary planning**		**10 Weeks**	**Mo 05.03.07**
2	Working out of fundamentals	Architect	1 Week	Mo 05.03.07
3	Preliminary planning project planning	Architect	2 Weeks	Mo 12.03.07
4	Preliminary planning engineering experts	Eng. expert	2 Weeks	Mo 26.03.07
5	Cost estimate facilities planning	Architect	2 Weeks	Mo 26.03.07
6	Cost estimate building services equipment	Eng. expert	2 Weeks	Mo 09.04.07
7	Compiling of preliminary planning	Architect	1 Week	Mo 23.04.07
8	Handover of preliminary planning to client	Architect	0 Weeks	Fr 27.04.07
9	Cost estimate + report preliminary planning	D&S	1 Week	Mo 30.04.07
10	Approval preliminary planning	Client	0 Weeks	Fr 11.05.07
11				
12	**Conceptual engineering**		**11 Weeks**	**Mo 30.04.07**
13	Preparation of first drafts	Architect	3 Weeks	Mo 30.04.07
14	Conceptual engineering by engineering experts	Eng. expert	3 Weeks	Mo 21.05.07
15	Preparation conceptual engineering	Architect	2 Weeks	Mo 11.06.07
16	Cost estimate facilities planning	Architect	2 Weeks	Mo 11.06.07
17	Cost estimate building services equipment	Eng. expert	2 Weeks	Mo 11.06.07
18	Compiling of conceptual engineering	Architect	1 Week	Mo 25.06.07
19	Handover of conceptual engineering to client	Architect	0 Weeks	Fr 29.06.07
20	Cost estimate + report conceptual engineering	D&S	2 Weeks	Mo 25.06.07
21	Approval conceptual engineering	Client	0 Weeks	Fr 13.07.07
22				
23	**Planning for permission to build**		**16 Weeks**	**Mo 02.07.07**
24	Preliminary coordination meeting with legal board of construction	Architect	1 Week	Mo 02.07.07
25	Preparation of submission plans	Architect	2 Weeks	Mo 02.07.07
26	Compiling of submission documents	Architect	0,5 Weeks	Mo 16.07.07
27	Signature of client on building application	Client	0,5 Weeks	Mi 18.07.07
28	Copying of submission documents	Architect	1 Week	Mo 23.07.07
29	Submission building application	Architect	0 Weeks	Fr 27.07.07
30	Review of building application	Building authority	12 Weeks	Mo 30.07.07
31	Granting of building permission	Building authority	0 Weeks	Fr 19.10.07
32				
33	Execution planning of shell construction	Archit./ Eng. exp.	35 Weeks	Mo 30.07.07
34	Execution planning of rest	Archit./ Eng. exp.	20 Weeks	Mo 07.01.08

Abb. 4–32 Excerpt timetable

The individual processes must offer themselves to be sorted according to trade and building section-specific criteria, since the various involved parties are otherwise confronted with too much information at once.
There are no rigid rules here; the desired ranking can be adjusted to the specific project and/or client.

Further ranking is possible according to fields of authority e.g. architect, expert engineering, support structure planners, builder etc. This way, everyone involved can see "at a glance" what tasks are to be completed when by him-or herself. This ranking then also forms the basis for schedule-related planning control through project management and can be equipped with remarks and agreements.

4.4.4 Gradual Structure of Systematics in Project Management

It is obvious that schedule statements become more and more inaccurate with an increasing distance to the occurrence of an event. For the preparation of the process structure plans, it must be taken into account that, at the beginning of a project, all details that could be important to the process are by far not known yet.

The statements concerning schedule processes, therefore, get all the more exact in respect to planning and advance of construction works the nearer the time of execution draws.

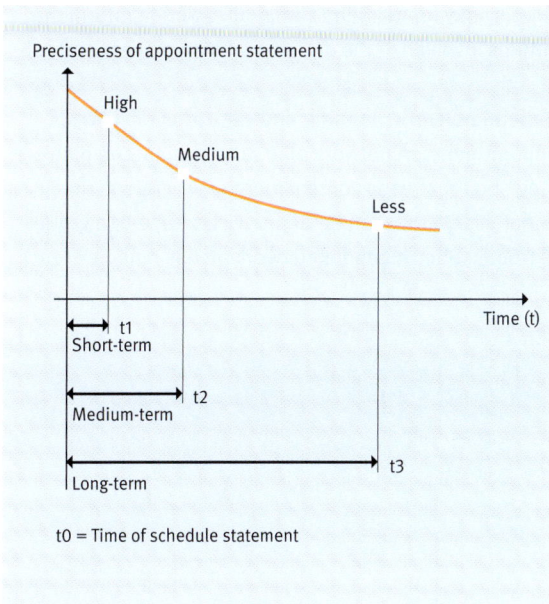

Fig. 4–33 Accuracy of schedule statements

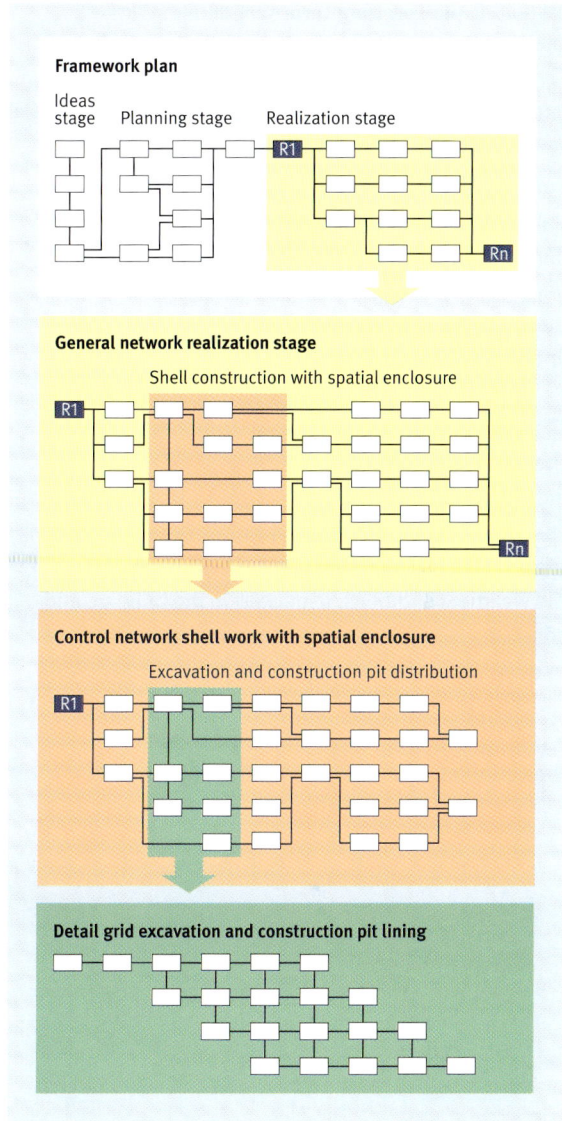

Abb. 4–34 Gradual structure of schedule planning

The requirement that a practice oriented time scheduling system must be built up gradually results from this realization. The individual stages, from the rough structure to the detailed plan, must be continuous.

The plans most important for the project manager are the outline plan or milestone plan and the general networks that he/she must create from knowledge of his/her own at a time when the "planned for" partners are, as a majority, not yet involved in the project at all. As a rule, the control nets can be created together with the persons concerned, or at least coordinated with these. The detail nets are built by the corresponding entities involved, checked by the project manager and taken over, in compressed version, for the control plans.

Milestone plan for long-term scheduling: With the assistance of the milestone plan, the complete time sequence of a project is fixed in its outlines. The individual events must be specified by empirical values in their temporal frame. The special difficulty when putting forward the milestone plan lies in the fact that many individual questions of the execution and implementation are not still known at this stage of planning. The analyses serving as a basis for the planning, hence, must be worked out as thoroughly and in as much detail as possible.

The described analyses allow for a good overview of the project. Specification of the execution deadlines alone does not suffice to render the construction process as optimal as possible. The interrelations between the individual work segments and the influence of the weather conditions, especially in the building and construction industry, must be taken into account and integrated into the plan.

It is particularly difficult during the development of the milestone plan to incorporate the phase when the

authorities need to be included in the time sequence for the purposes of granting of the planning and construction permission. These do not like to be pushed into committing to specific deadlines. Experience, however, shows that this phase, too, can be brought under control through prior clarification with the authorities concerned.

The inclusion of process-defining delivery times is important already in the milestone plan, if these exceed the usual three to six months. Delivery dates from one to two years can sometimes arise, particularly in the context of special machines or electrical engineering equipment for industrial construction. It is the task of project control, through early queries and obtaining information, to find out about such points as are particularly risky for deadline compliance and to inform the entities involved in the decision-making process early on.

General network planning and execution of contract appointments: A network plan for medium-term planning should already contain more exact statements about events that can be defined by a trade or a group of trades. Such a general network plan is built in

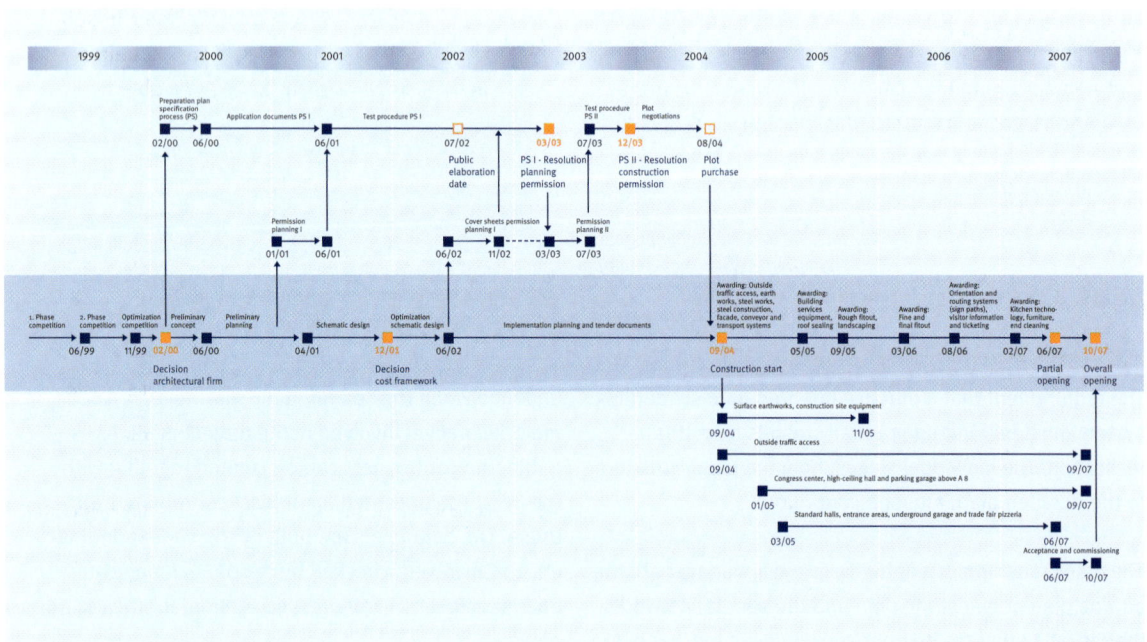

Fig. 4–35 Milestone plan as network plan

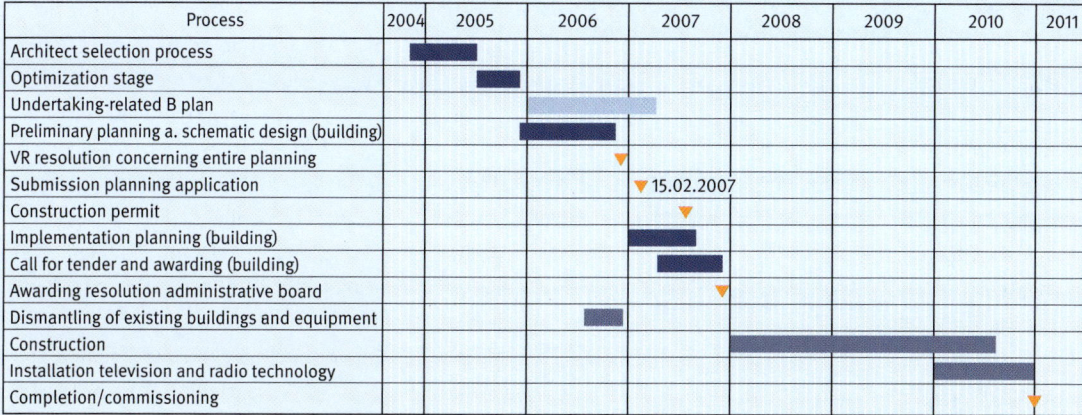

Process	2004	2005	2006	2007	2008	2009	2010	2011
Architect selection process								
Optimization stage								
Undertaking-related B plan								
Preliminary planning a. schematic design (building)								
VR resolution concerning entire planning								
Submission planning application				15.02.2007				
Construction permit								
Implementation planning (building)								
Call for tender and awarding (building)								
Awarding resolution administrative board								
Dismantling of existing buildings and equipment								
Construction								
Installation television and radio technology								
Completion/commissioning								

Fig. 4–36 Milestone plan as gant chart

block- or sub-networks which are connected to each other. An essential task of the general network is the determination of contract deadlines.

To be able to define these contract deadlines for certain, the essential individual processes of the company concerned, its predecessors and its successors must be included in the general network plan. Building engineering can be regarded as an example for such a sub-network. Planning lead times and delivery times are included, as far as important service components are concerned. In order to be able to immediately identify the essential dependencies, on the predecessors and the effects on the companies to follow, of an amendment of the contract deadlines, representation as a network plan, better still as an intermeshed gant chart, is required.

Especially when penalties for breach of contract have been agreed on or the contractor has to pay for subsequent costs that result from his/her schedule delays, such a general network plan is indispensable.

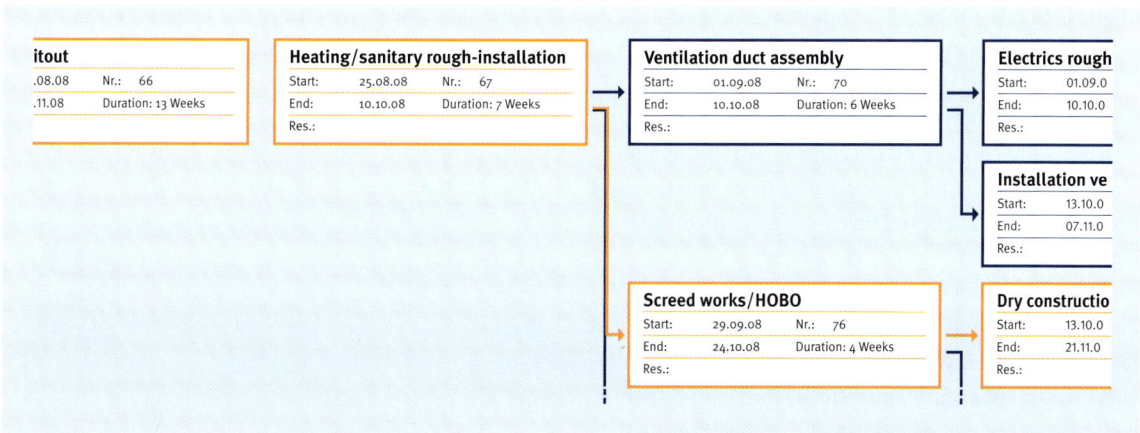

Fig. 4–37 Contract schedule building engineering

Control plans: Control planning serves for direct implementation of deadline specifications with those involved in the project.

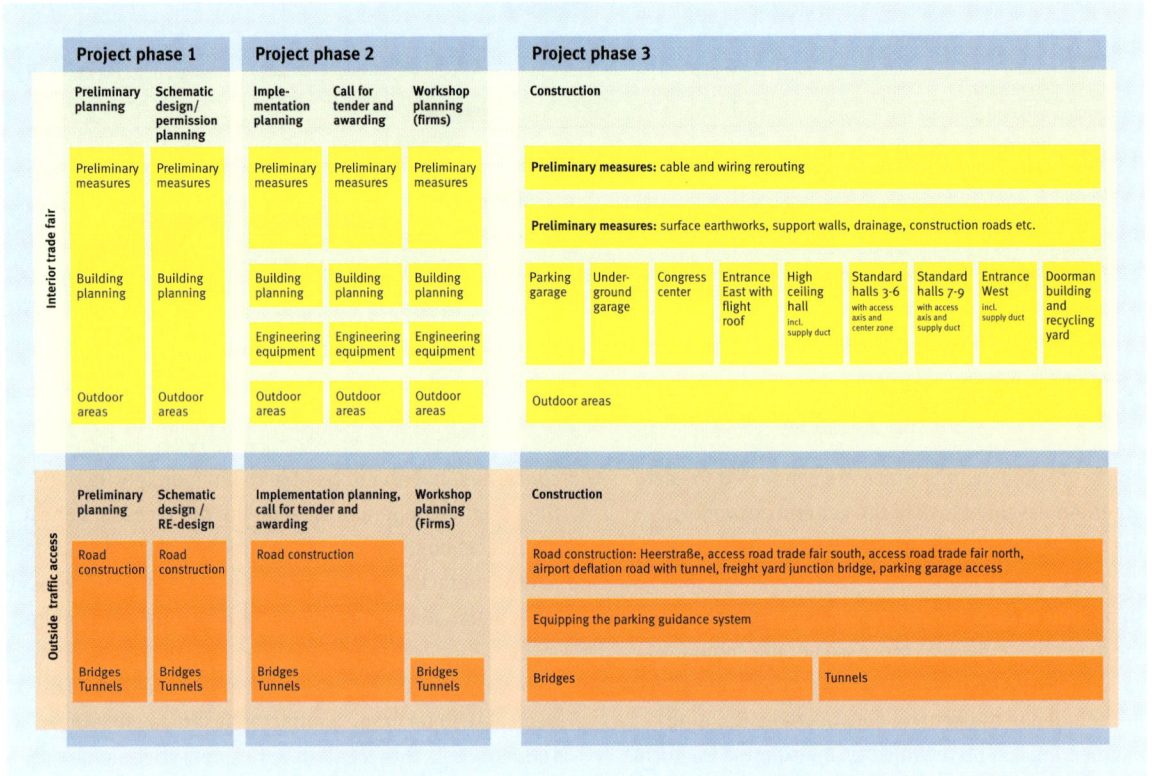

Project phase 1		Project phase 2			Project phase 3
Preliminary planning	Schematic design/ permission planning	Imple-mentation planning	Call for tender and awarding	Workshop planning (firms)	Construction

Interior trade fair

Preliminary measures	Preliminary measures	Preliminary measures	Preliminary measures	Preliminary measures	**Preliminary measures:** cable and wiring rerouting
					Preliminary measures: surface earthworks, support walls, drainage, construction roads etc.
Building planning	Building planning	Building planning	Building planning	Building planning	Parking garage / Under-ground garage / Congress center / Entrance East with flight roof / High ceiling hall incl. supply duct / Standard halls 3-6 with access axis and center zone / Standard halls 7-9 with access axis and supply duct / Entrance West incl. supply duct / Doorman building and recycling yard
		Engineering equipment	Engineering equipment	Engineering equipment	
Outdoor areas	Outdoor areas	Outdoor areas	Outdoor areas	Outdoor areas	Outdoor areas

Outside traffic access

Preliminary planning	Schematic design / RE-design	Implementation planning, call for tender and awarding		Workshop planning (Firms)	Construction
Road construction	Road construction	Road construction			Road construction: Heerstraße, access road trade fair south, access road trade fair north, airport deflation road with tunnel, freight yard junction bridge, parking garage access
					Equipping the parking guidance system
Bridges Tunnels	Bridges Tunnels	Bridges Tunnels		Bridges Tunnels	Bridges / Tunnels

Fig:. 4–38 Structure of control schedule planning

Control plans planning of the planning process:
In essence, planning is controlled by timetables.

An exact calculation of plan lead times and their schedule-and-content related control, with the aim of safe plan supply, is an indispensable basis for execution according to schedule and this requires particularly experienced staff.

Plan lead times must be coordinated with all involved in order to arrive at a sure basis for planning control. This is of greatest importance since more than 70 per cent of the schedule delays usually can be attributed to delayed plan deliveries. Causes can be the architect and the client because of changes, the authorities through stipulations they issue, or the planners for capacity reasons.

Plan delivery list shell work building section A1 and A2

Control date: 19.10.09   Basis: Implementation schedule SP-ARC

Level	Plan segment	Planner in charge	Envisioned construction start	Plan lead time [in weeks]	Plan delivery deadline – target	Plan delivery deadline – actual	Done (Yes)
General documents	View onto the roof (VR)	SP-ARC	01.12.08	4,0	03.11.08		
	Building cross-section	SP-ARC	01.12.08	4,0	03.11.08		
	Building views	SP-ARC	01.12.08	4,0	03.11.08		
	Outdoors facility plan (VA)	SP-ARC	01.12.08	4,0	03.11.08		
	Static calculations for pre-assembled components (Basis for company planning)	SP-ARC	01.12.08	4,0	03.11.08		
Excavation, Foundation, drainage, Floor slab	Overview plan construction pit securing	SP-ARC	01.12.08	4,0	03.11.08		
	Drainage plans, concrete footing ground, drainage planning	SP-ARC	01.12.08	4,0	03.11.08		
	Coordinated workshop plan Level-1 incl. detail cross-sections/specifications (Approved for constr.)	SP-ARC	01.12.08	4,0	03.11.08		
	Approved formwork plan foundation/floor slab (Approved for construction)	SP-ARC	01.12.08	4,0	03.11.08		
	Approved reinforcement plans foundation/floor slab (Approved for construction)	SP-ARC	01.12.08	4,0	03.11.08		
	Approved for construction reinforcement steel item lists + material excerpts foundation/floor slab	SP-ARC	01.12.08	4,0	03.11.08		
	Approved for construction emtpy duct plans for electrical installations level -1	SP-ARC	01.12.08	4,0	03.11.08		
	Specific. for installation components of elevators, building serv. equipm. + facade found. /floor slab	SP-ARC	01.12.08	4,0	03.11.08		
Level -1	Coordinated workshop plan level 0 incl. detail cross-sections/specifications (Approved for constru...	SP-ARC	01.12.08	4,0	03.11.08		
	Approved formwork plan ceiling above level -1 (Approved for construction)	SP-ARC	01.12.08	4,0	03.11.08		
	Approved reinforcement plans ceiling above level -1 (Approved for construction)	SP-ARC	01.12.08	4,0	03.11.08		
	Approved for construction reinforcement steel item lists and material excerpts ceiling above level 1	SP-ARC	01.12.08	4,0	03.11.08		
	Approved for construction empty duct plans for electrical installations level -0	SP-ARC	01.12.08	4,0	03.11.08		
	Specific. for installation components of elevators, building services equipment and facade level 1	SP-ARC	01.12.08	4,0	03.11.08		
Level 0	Coordinated workshop plan Level-1 incl. detail cross-sections/specifications (Approved for constru...	SP-ARC	16.03.09	8,0	19.01.09		
	Approved formwork plan ceiling above level -0 (Approved for construction)	SP-ARC	16.03.09	8,0	19.01.09		
	Approved reinforcement plans ceiling above level -0 (Approved for construction)	SP-ARC	16.03.09	8,0	19.01.09		
	Approved for construction reinforcement steel item lists and material excerpts ceiling above level 0	SP-ARC	16.03.09	8,0	19.01.09		
	Approved for construction empty duct plans for electrical installations level -0	SP-ARC	16.03.09	8,0	19.01.09		
	Specific. for installation components of elevators, building services equipment and facade level 0	SP-ARC	16.03.09	8,0	19.01.09		

Fig. 4-39 Plan delivery lists

Control plans construction: The general network plan and the updated process structure plans form the basis for the control network plans. These are created for the schedules, on the level of the short-term appointment supervision and control. For these plans, details ought to be thus advanced that all events belonging to the various fields of responsibility or making use of the different factors of production as represented one by one.

Nr	Process name	Responsible	Schedule (Jul '08 – Jan '09)
64	**Rough fitout**		25.08. ... Rough fitout
65	**Basement**		25.08. ... Basement
66	Heating/sanitary rough-installation	H/S	25.08. 05.09.
67	Installation heating/sanitary control room	H/S	13.10. 07.11.
68	Building heating possible from	H/S	07.11.
69	Ventilation duct installations	V	25.08. 05.09.
70	Rough installation, electrics	E	25.08. 12.09.
71	Installations LVM (Low Voltage Mains)	E	15.09. 03.10.
72	Installation steel doors	Dry constr. works	25.08. 05.09.
73	Locksmith part 1	Locksmith	08.09 12.09.
74	Screed works/HOBO	Screed	15.09 19.09.
75	**Ground floor**		25.08. Ground floor
76	Heating/sanitary rough-installation	H/S	25.08. 05.09.
77	Ventilation duct installation	V	25.08. 05.09.
78	Rough installation, electrics	E	25.08. 05.09.
79	Wet plaster	Wet plaster	08.09. 19.09.
80	Locksmith part 1	Locksmith	22.09. 26.09.
81	Screed works/HOBO	Screed	29.09. 10.10.
82	Dry construction work incl. door frames	Dry constr. works	13.10. 24.10.
83	**1. floor**		08.09. 1. Floor
84	Heating/sanitary rough-installation	H/S	08.09. 19.09.
85	Ventilation duct installations	V	08.09. 19.09.
86	Rough installation, electrics	E	08.09. 19.09.
87	Wet plaster	Wet plaster	22.09. 03.10.
88	Locksmith part 1	Locksmith	06.10. 10.10.
89	Screed works/HOBO	Screed	13.10. 24.10.
90	Dry construction work incl. door frames	Dry constr. works	27.10. 07.11.

Fig. 4-40 Control plan construction

Fine planning for an area far away the planning point does not make sense. This means that, along with the project progress, planning is also refined e.g., the control plans are only in advance of the respective execution time by around a certain time interval. As a rule, one will refine and unanimously adopt certain sub-networks, such as shell construction procedure, in the context of the awarding of the work in cooperation with the enterprises receiving the contract. The control plans of later phases, therefore, build respectively on the specifications of the general net and on the coordinated advance performances. So the control networks should be detailed in only such a manner as is compatible with the aims that are to be achieved with this network plan.

Prefailed network- or gant charts: With increasing fine-tuning degree of planning, planning effort rises excessively. In the execution stage, the interest of the persons responsible declines noticeably if the responsibility leeway where they can act on their own is restricted too strongly for them by too detailed a planning.

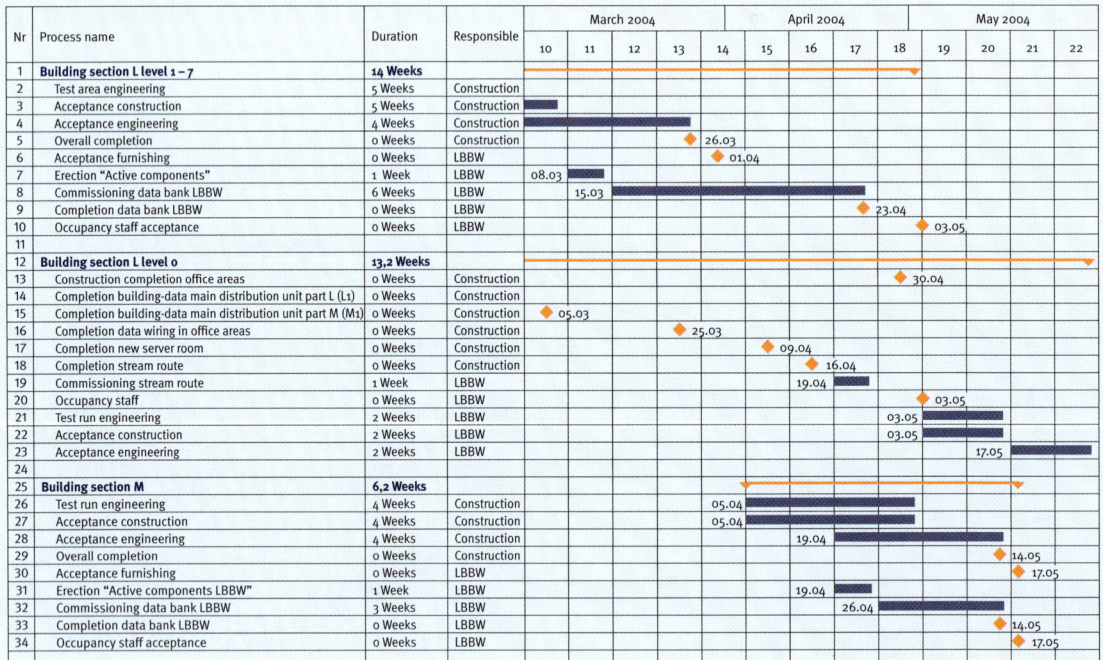

For particularly critical areas, such as excavation pit lining and excavation, it can become necessary to introduce another detail step together with the detail net. Since this detail net already is on the level of work preparation and/or planning, it ought to either be created jointly with the executing company or else co-ordinated intensively with the same.

Another main application area for detailed schedule plans is the phase of commissioning.

Here, very detailed control must be carried out, section and area-wise, in connection with the removal of any shortcomings – as represented in Fig. 4–41.

Nr	Process name	Duration	Responsible
1	**Building section L level 1 – 7**	**14 Weeks**	
2	Test area engineering	5 Weeks	Construction
3	Acceptance construction	5 Weeks	Construction
4	Acceptance engineering	4 Weeks	Construction
5	Overall completion	0 Weeks	Construction
6	Acceptance furnishing	0 Weeks	LBBW
7	Erection "Active components"	1 Week	LBBW
8	Commissioning data bank LBBW	6 Weeks	LBBW
9	Completion data bank LBBW	0 Weeks	LBBW
10	Occupancy staff acceptance	0 Weeks	LBBW
11			
12	**Building section L level 0**	**13,2 Weeks**	
13	Construction completion office areas	0 Weeks	Construction
14	Completion building-data main distribution unit part L (L1)	0 Weeks	Construction
15	Completion building-data main distribution unit part M (M1)	0 Weeks	Construction
16	Completion data wiring in office areas	0 Weeks	Construction
17	Completion new server room	0 Weeks	Construction
18	Completion stream route	0 Weeks	Construction
19	Commissioning stream route	1 Week	LBBW
20	Occupancy staff	0 Weeks	LBBW
21	Test run engineering	2 Weeks	LBBW
22	Acceptance construction	2 Weeks	LBBW
23	Acceptance engineering	2 Weeks	LBBW
24			
25	**Building section M**	**6,2 Weeks**	
26	Test run engineering	4 Weeks	Construction
27	Acceptance construction	4 Weeks	Construction
28	Acceptance engineering	4 Weeks	Construction
29	Overall completion	0 Weeks	Construction
30	Acceptance furnishing	0 Weeks	LBBW
31	Erection "Active components LBBW"	1 Week	LBBW
32	Commissioning data bank LBBW	3 Weeks	LBBW
33	Completion data bank LBBW	0 Weeks	LBBW
34	Occupancy staff acceptance	0 Weeks	LBBW

Abb. 4–41 Detail plan commissioning

Use of time-route diagrams for skyscrapers: The organization and time scheduling for skyscrapers is essentially characterized by logistics, owing to the unusual feature of a vertical line construction site. A number of skyscraper-specific scheduling factors of influence also have to be taken into account already before construction start, however.

In principle, we can assume that a short overall construction time can only be achieved by optimization and coordination of all individual events. It must be considered that, during skyscraper construction, special significance is placed, due to linear vertical extension, on drafting of interval methods.

For the time-route diagram, the path to be taken is determined by the building itself. For this, we transfer a cross-section of the building to the plan, to which we then add a time axis. The next step is then to deter-mine sequences for shell construction and facade. The ascent of the lines, here, indicate construction speed. In the example at hand, the facade starts from the first floor. Facade assembly follows shell work at the same speed, meaning that optimum conditions are provided for sub-sequent installation of engineering and fitout.

During shell work, the typical S-curve can be identified. It corresponds initially to a lower construction speed for the basements, speeding up for the regular floors and, as a rule, receding again for the upper floor region. Based on shell work and facade, we can then start scheduling – shown here in a very simplified manner – in-house engineering and fitout works.

In-house engineering, here, needs to be adjusted to the adherence of a certain minimum distance to the facade as well as the prerequisite of a horizontal sealing. The fitout works can then follow the in-house engineering installation, floor-by-floor. Basements and first floor are usually scheduled, since there are other functional connections there. As a rule, technology assembly in the basements can start once the ceiling over the first floor has been completed. Works in the basement ought to be undertaken only after completion of the facade.

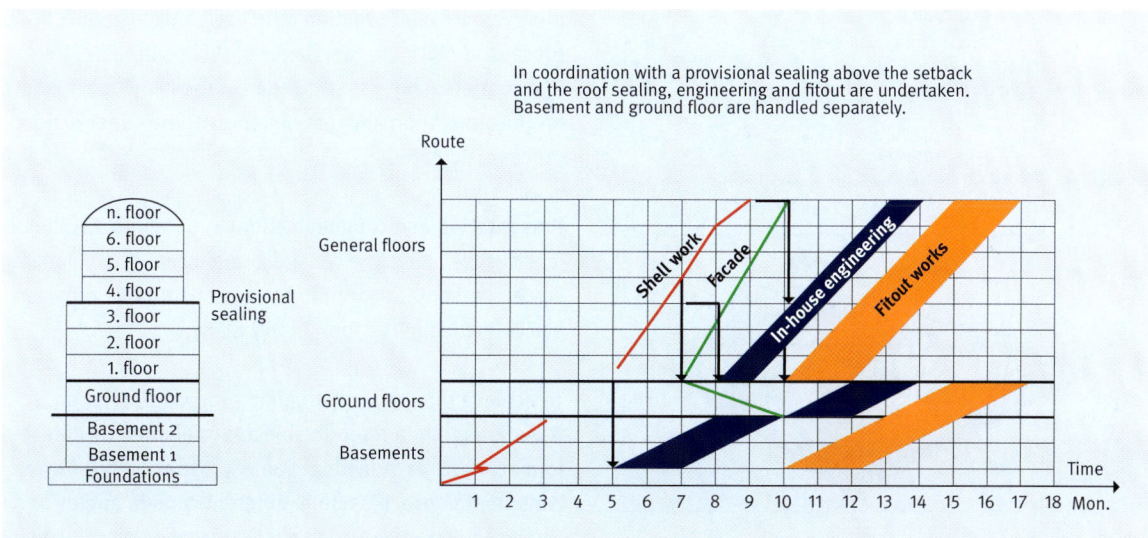

Fig. 4–42 Time-route-scheduling diagram

4.4.5 Construction Process Simulations

It requires efficient planning and management tools to realize short-time but complex construction projects on schedule and without additional costs. The tools condense and abstract construction processes and their mutual dependences. A construction process simulation is important also for experienced construction experts, for early identification of problem areas on the construction site in the midst of an array of diagrams and network plans. For very complex construction undertakings, the possible problem areas can made visible sooner and sequential alternatives suggested. Even without expert knowledge, all committees involved have a clear decision basis through this.

Procedure: To be able to simulate the progress of works three-dimensionally, existing plan documents, like site plan, ground plans, cross-sections and views, as well as additional details on the stages of construction, construction site facilities and photos of the given situation are used.

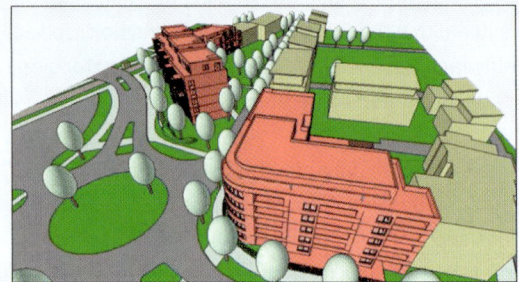

Fig. 4–43b 3 D animation of the construction process

From these sources, a structured data model of the project and its direct surroundings is created. A 3 D model results that corresponds to the real model at a scale of 1:500.

For the process simulation control file of the construction progress, time information from the frame schedule is processed. With the assistance of this control file, the data model for the 3 D calculation of the individual frames for the simulation and/or animation of the construction process is created.

Presentation and communication: In principle, the client receives the simulated construction process as a computer screen presentation. Additionally, construction progress can be issued in the form of DIN plans.

To achieve this, the construction process simulation is transferred into a reproducible and scalable PDF data format. 25 to 30 individual frames, respectively, show construction progress from different camera angles and can be equipped with additional explanations and legends.

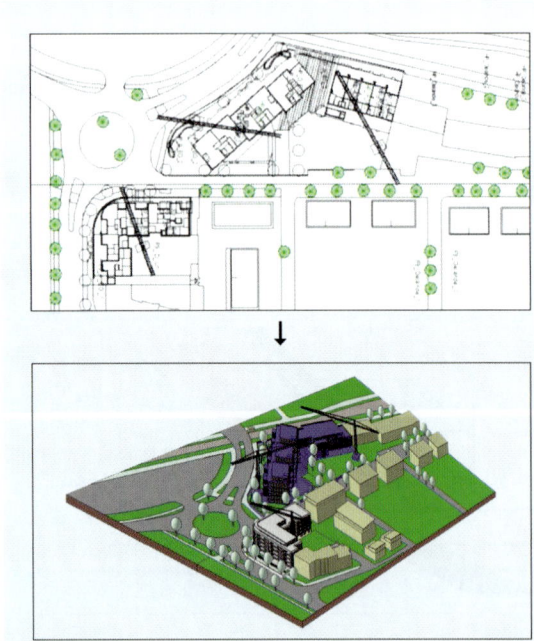

Fig. 4–43a Creation of a 3 D model

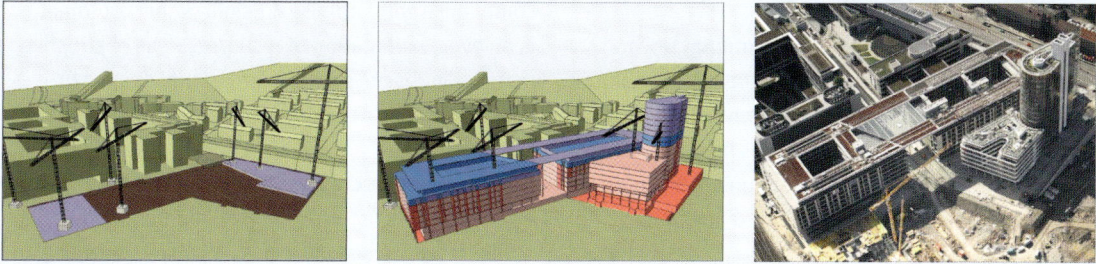

Fig. 4–44 Construction process simulation of an administrative building (© Drees & Sommer)

For the preparations for complex building schemes, the 3 D simulation is a particularly well-suited instrument because it spatially displays the manner in which the process of the construction undertaking has been planned. Construction process simulation is an important basis for optimized and stable project preparation and planning. It clarifies temporal and local contexts. Since actual planning data is integrated for this, the visualization provides a realistic image of future construction activity but avoids difficult to "read" abstraction of schedules and diagrams. The possibility of having the construction site shown from different perspectives as well as the vivid manner in which the information is provided simplify and shorten processes of coordination.

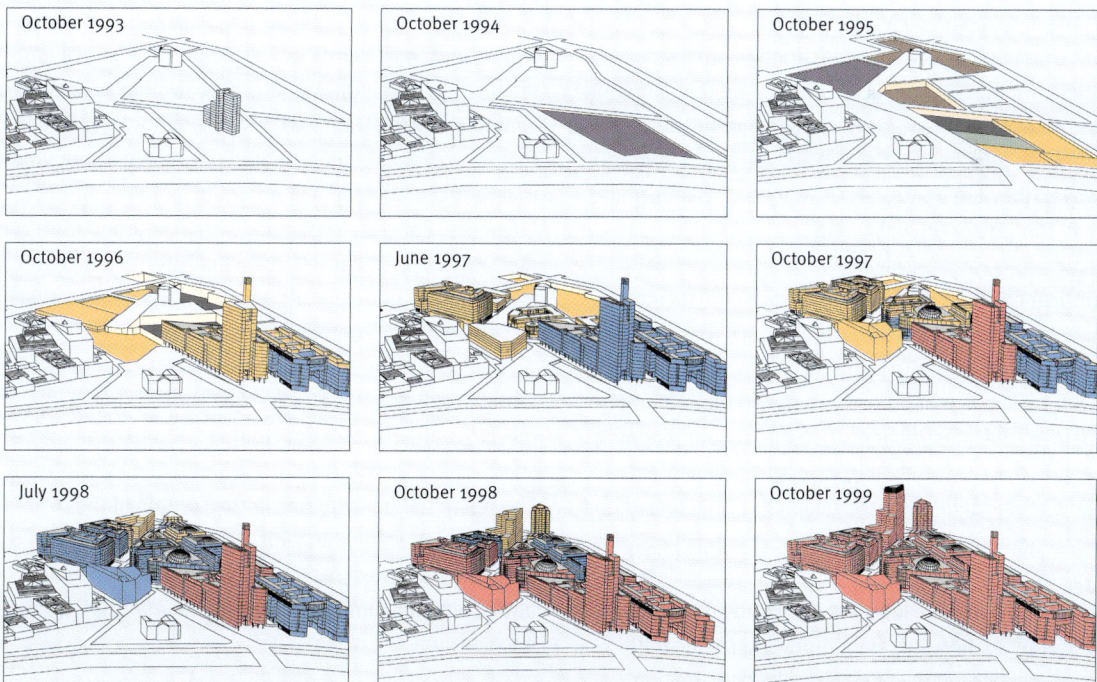

Fig. 4–45 Construction process simulation Potsdamer Platz Berlin (© Drees & Sommer)

4.4.6 Logistics Planning and Construction Site Facilities

Demands on construction logistics planning: Downtimes are an expensive luxury. Only if the construction site processes smoothly overlap can productivity and economical aspects of construction sites be considerably improved. Downtimes and improvisation are a luxury that a construction project cannot afford anymore.

It is all about using employees, equipment and material in right place, time and in the right amount while keeping an eye on the costs for procurement, transportation and storage at the same time. This offers the best starting-points for developing savings potentials, increasing economical aspects and productivity on the construction sites and create a completely new information quality.

Construction operation and project management must work together in the interest of continuous and efficient processes. Building economically, among other things, means to avoid idle times because of missing equipment or due to material bottlenecks. Logistics costs frequently have a stronger impact on the acquisition of materials than the actual purchase price. Optimal use of the scarce goods, area and time at the construction site is a prerequisite for functioning, is the enforcement of common rules of the game. Here, sanction possibilities are agreed on between principal and agent, and these are important. However, they are only rarely used if the logistics planning is carried out professionally.

Logistical concept: Project management must create a concept for the construction site facilities plan already in the planning phase, in order to incorporate the know-how of efficient processes. Like in the stationary industries, the construction site gets a layout planning related to the construction phase. Specifying this early on saves costs and shows the building contractors under which conditions they can design their respective trades.

However, not only the logistics have to be coordinated on the construction site; neighbors and public traffic also have claims on the often stressful construction time. Frequently, acceptance and tolerance among the persons affected can be obtained by a few means,

communication, qualified control of the processes and compliance with promises. Result of this coordination is an even exploitation of resources, less obstructive material in rescue paths, less material decline or damages to and accidents at the construction site.

Logistics for skyscrapers: Since skyscrapers are essentially erected within congested urban areas, as a rule the construction field is calculated only just. This, in principle, leads to the need of having external ware-housing with composite parts manufacture. If possible, there should be just in time on the spot assembly for all areas, e.g. all construction and auxiliary construction materials should be delivered so that they can be processed within shortest time. This is aggravated again by the fact that a skyscraper virtually represents a vertical line construction site, which allows only a very restricted influence of means of transportation due to the low base area that is available. So the aim is a minimization and optimization of transportation events, something that in turn strengthens the demand for outside far-reaching preassembly of parts. The construction time plan must be created in coordination with a detailed transportation schedule, which, in turn, requires exact planning of the available means of transportation. As a rule, two high-performance climbing cranes, from 250 to 500, are nowadays used that, however, generally should only serve for transport of reinforcements, ceiling boarding as well as facade elements. Concrete transport, with or without intermediate stops, is carried out via concrete pumps and the formwork of core walls and outer walls is installed as self-climbing formwork to relieve the crane.

Corresponding outer elevators as well as co-growing elevators are used for development in the core areas.

The restricted possibilities in the context of the overall logistics, in connection with the demand for short construction times, require an early look at the shell and extension concepts and their coordination with each other.

Fall 2004: Preliminary measures

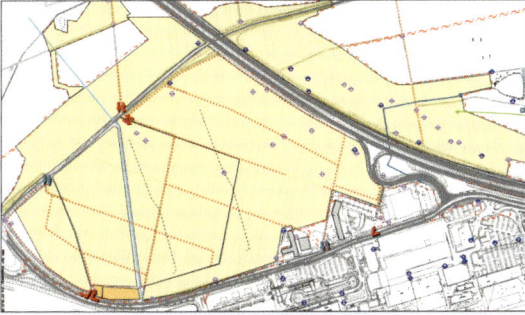

Summer 2005: Earthworks and shell

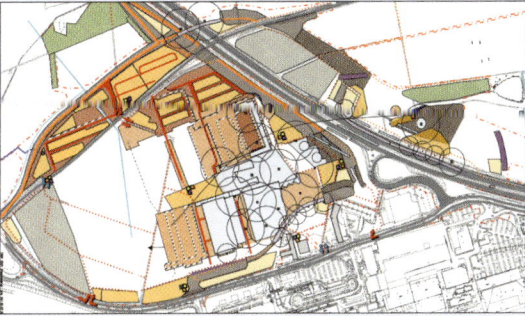

Spring 2006: Shell and spatial enclosure

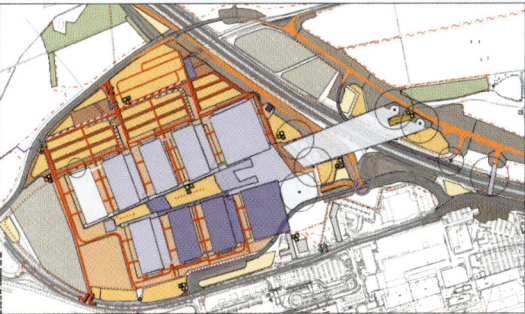

Summer 2007: Fitout and completion

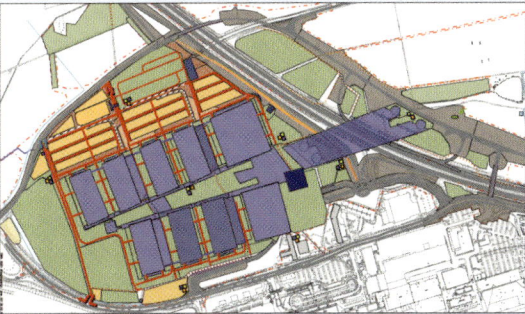

Fig. 4–46 Logistics concept and implementation in practice

Safe disposal: Since 2005, disposal costs on construction sites have increased by over 100% through the new Technical Instructions Settlement Waste (TISW); looked at more closely, however, this only applies to mixed waste prices. Almost 75 % of total costs for disposal are incurred in waste transit and only 25 % for processing. Rolling containers or assembly bags are therefore recommend, which considerably reduce transport costs. Through this, every executing company can already separate the waste at the processing stages and throw it into the lockable containers, or the insulation materials into the empty assembly bags. Alternatively, the waste should be thrown on the floor again prior to transport, and be separated and taken away by the cleaners.

Fig. 4–47 Assembly bags for insulation work on the scaffolding

The clients are also content, aside from the executing companies, since the process has a number of positive side effects:

– A cleaner construction site
– Less vertical transports owing to rolling containers
– Higher productivity
– Less danger of fire
– Less quarrellings over waste origin

Security and checks on people: The fight against illegality plays a special role in the construction trade. The hope for a cheap awarding of a service has turned out to be a fallacy: Often, unexpectedly high costs have resulted – far greater than would have been incurred by checking out the people working on the site - through demands for back payment of national insurance contributions by the clients or the damage done by unintentional publicity in the press after a raid. Not only does site access control prevent illegal employment but it also reduces theft and damage. The staff employed for this task not only monitor entrance to the construction site but also the outer appearance of the site – an important indication for the realization of a high-quality property.

Safety and health coordination: Project management must guarantee all services in connection with the requirements of the construction site ordinance. This can be carried out relatively economically in interplay with logistics planning and supervision:

– Through access control, all new enterprises, their numerous sub-entrepreneurs as well as entrepreneurs without employees or employers who are active themselves on the construction site, are registered and can be instructed specifically.
– First helpers also are included in the control system and through this, sufficient presence on the construction site is also under supervision.
– The logistical coordination of the delivery and storage of material reduces amounts present at the site and thus minimizes the blocking of escape routes.
– The disposal system with rolling containers described above, for fitout stages, reduces the dangers from offcuts and waste lying around.
– For the supervision of the cleaning work, the daily required shortcomings management reduces the danger of stumbling and unnecessary fire hazards on the floors.
– These logistics coordinators present on the construction site are in constant with the relevant authorities and thus are able to react swiftly.

4.4.7 Process Control (Control Cycle Schedule Control)

Even when all requirements for optimum project implementation have been met like

- Effective project organization,
- Efficient planning of processes and
- Well thought-out logistics planning

there is still a remainder of interfering factors that cannot, or only with difficulty, be removed by the above-mentioned measures. Unexpectedly lengthy bad weather periods, failure of important trades through bankruptcy or similar events force a particularly elastic approach in the construction execution phase. One makes this possible best by getting a clear picture, in regard to future crash case considerations, about possible alternatives.

In any event, primarily in the case of extremely short deadlines, so-called calm-down phases should be planned for, in which any appearing delays can be dealt with – despite everything.

The following points can considered as essential prerequisites for a successful sequence control:

- The process planned with the assistance of the network planning method must be practicable – e.g., it must have been created by experienced staff.
- The necessary information must be transmitted to the people involved – in a clear and understandable manner.
- The project manager must have so much experience and authority that his orders are also then accepted when they are not regarded as optimal by all involved.
- The project manager must not "stop those involved from doing their work" – e.g., control must be organized in such a manner that he or she should annoy them as little as possible (no endless meetings, particularly with people who are not even involved).

Control of the planning: In the area of the planning, project management deals with a relatively small number of planning leaders so that direct contact makes sense here between planners and project manager.

Appointment control has to be carried out in the planning area with the help of feedback lists that allow project management to identify deviations from target deadlines and to record the same.

A coordination meeting takes place on the basis of these protocols, in a 2 till 3-week rotation. During this meeting, discrepancies are immediately cleared away and measures specified to deal with delays.

Control of project execution: During project execution, the number of those involved rises dramatically so that a constant contact of project management to all involved does not make sense here any more. The following sequence has therefore taken shape for control of execution:

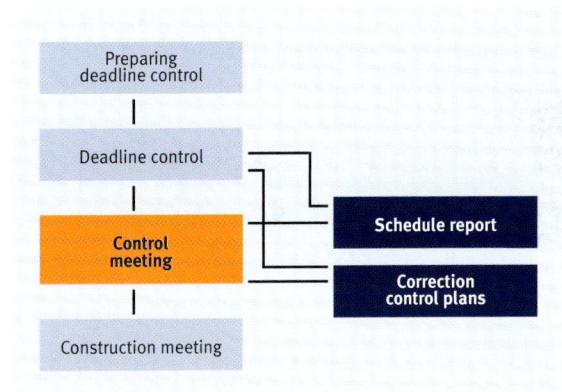

Fig. 4–48a Control meeting

Preparing deadline control: By the specified day, the project manager arranges all target data to be checked against a target-is list. In addition, the required process plans are processed visually as a basis for discussion for deadline control and the control meeting itself.

Deadline control: Deadline control on the construction site is undertaken by the project manager. From deadline control, all events that do not agree with the target dates are filtered out now. It is checked for these events whether they lie within an appointment buffer or not.

Control meeting: In the control meeting, critical events are now talked over, using the deadline protocol. Suitable measures that may assist in dealing with the delay are to be suggested by the project manager and other involved parties. These suggestions, for instance, may involve:

– Capacity increases
– Shortening of subsequent events
– Changes to process structure
– Special measures (winter construction, provisional arrangements etc.)

For larger projects, the required measures must already be in a state to be decided on, least up to a certain order of magnitude, in the control meeting. This requires the presence of a competent representative of the client during this control meeting. As for the rest, the architect and the specialized engineers as well as possibly also some companies, as required, are to be consulted.

Schedule report: From the results of schedule control and the control meeting, a short report has to be compiled for the client by project management. From this report, the following ought to result:

– Required decisions of the client
– Status of the work (overview)
– Special occurrences
– Special measures (primarily if these are connected with additional costs)
– Forecast for deadline compliance

The report should not exceed the size of two to three pages and be sectioned clearly.

Correction of the control plans: The control plans must be internally corrected after each schedule control. Specification of the corrected deadlines is undertaken by the issue of appointment lists at 6-week rhythm. These lists need to include the target data for the next six weeks.

Construction meeting: In the construction meeting, the area and specialist construction building managers must succeed against the companies with the control measures decided on. In serious cases, overriding construction managers, project managers and planners also are consulted for the construction meeting.

Fig. 4–48b Control meeting large-scale project

4.5 Cost Management

A client, in general, is dependent on costs for a building being kept to a minimum in the context of the defined aims. Hence, market rents to be expected in a certain range do not allow any arbitrary project costs, but rather a maximum budget which must be adhered to, despite an as good as possible marketing-promoting equipment. To achieve this, four requirements have to be observed:

– An **obligatory cost scaffolding**, which also makes a comparison with other projects possible, needs to be used. The cost scaffolding must not change during the entire project execution, so that control is possible for the highest aggregation level at any time.

– The budget must carefully be investigated, fault-free and exact. It must be absolutely clearly and obviously defined; what is not contained in the budget and what is. **The client must know all costs that he or she has coming to him or her.**

– All **activities** must correspond to the respective planning or realization stage.

– The **project cost control system** of the construction manager must always – via suitable interfaces – supply the superseding data through a translation program adapted to the type of structure the respective client wishes to receive.

The possibilities of influencing a construction project are largely exhausted by the end of the planning phase. For this reason, careful cost planning, especially in the early project phases, is very important.

The challenge lies in that only little information about the undertaking is available in the first project phases and from these, however, cost statements with a high precision degree are required. Since, with advancing project course, the influencing possibility decreases strongly, all required decisions must be undertaken during the initial project phases.

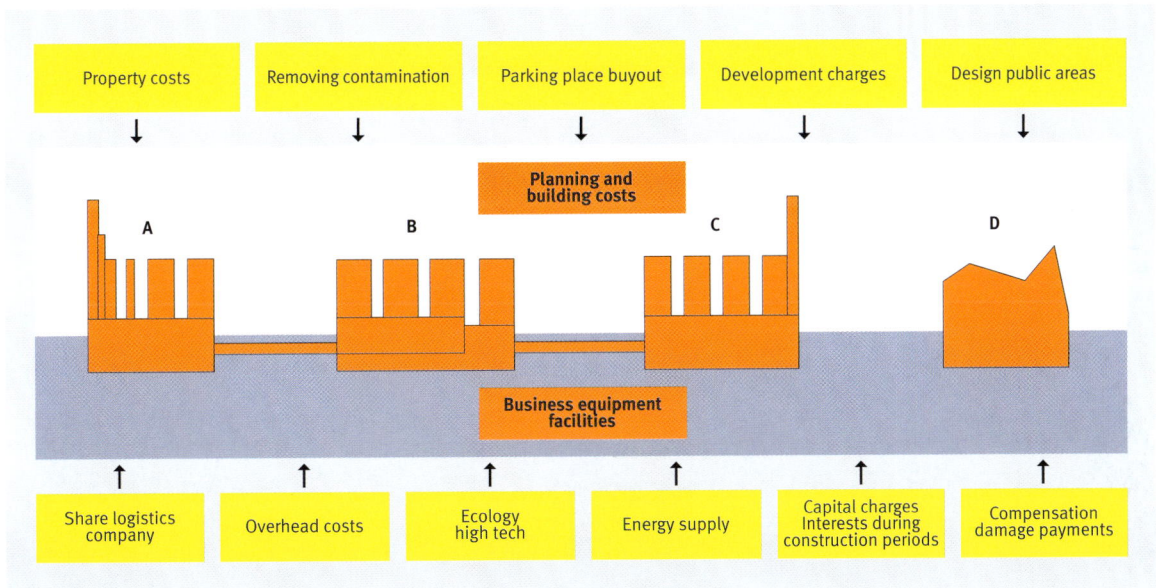

Fig. 4–49 Complete cost scaffolding

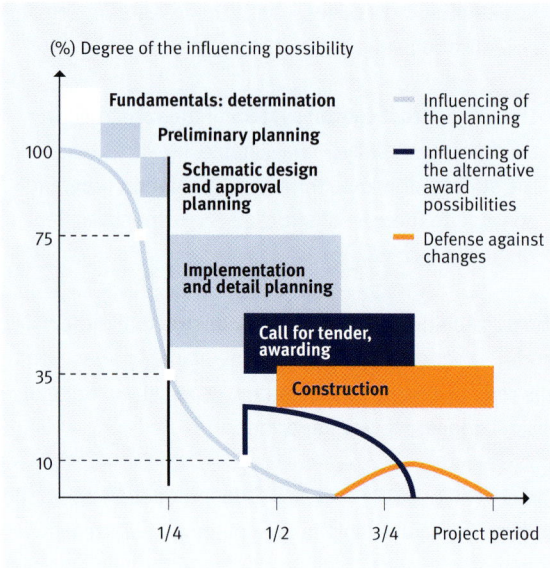

Fig. 4-50 Influencing possibility concerning the project course

The determination of the planning costs and the mechanisms for compliance to the same, can be excellently explained by depiction as a control cycle. The principle, here, works similarly to that of a thermostat valve for heating.

Once the budgeted costs have been defined, one has adjusted the "target temperature" so to speak. With a temperature sensor, the actual or Is temperature is measured and compared to the target. If there are any deviations, the thermostat adjusts the valve. It is similar for cost deviations: If additional costs are registered, then there are different possibilities for counter-control. If the costs remain below target, the client could fulfill a special wish for him or herself, provided that he hears about the cost reductions in time.

This makes clear that, during the execution decisions and with that the specification of the budgeted costs, there is a decisive focus on the course taken during the first planning phases while, as of execution planning, cost supervision becomes increasingly more meaningful.

Deviations from the target temperature arise from disturbances, such as fluctuations in outside temperature or an open door. At cost control, such fluctuations are cyclical or caused by planning or standard changes. Early diagnosis is decisive, for which a clear basis (cost planning) and a suitable control system (cost supervision) are required.

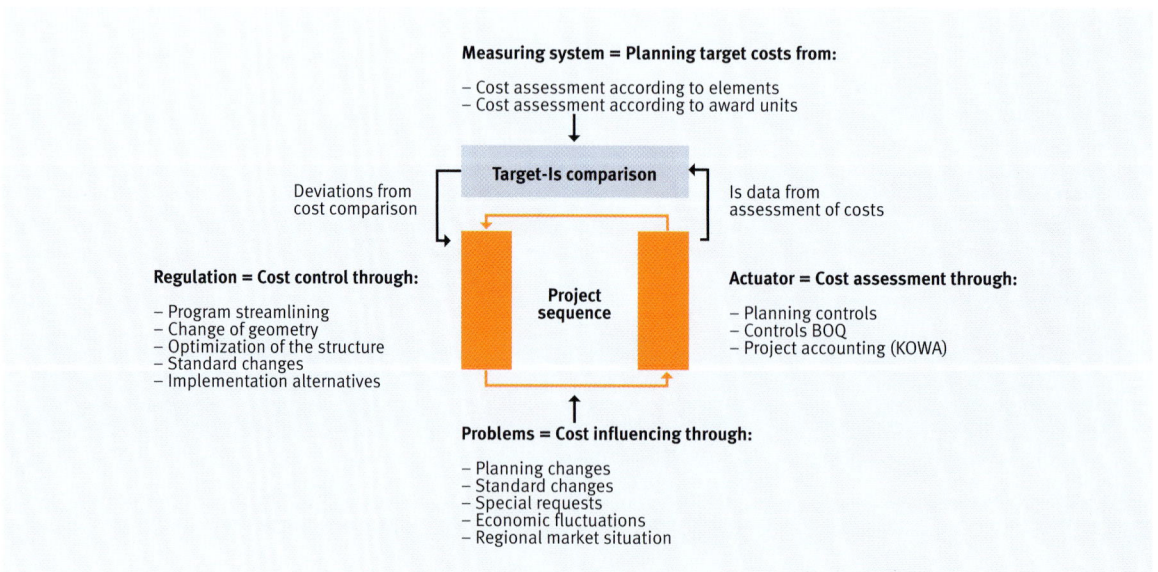

Fig. 4-51 Cost control as a control cycle

To be able to realize these requirements, cost management is structured into two performance phases.

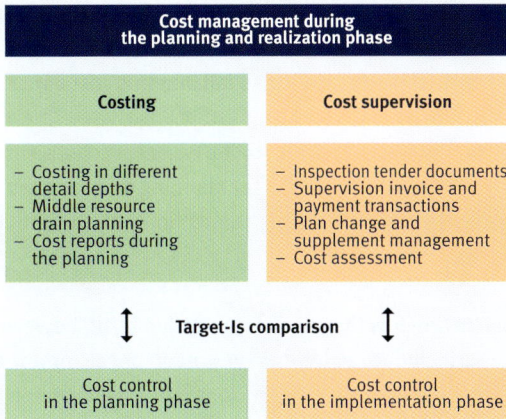

Fig. 4-52 Performance phases cost management

4.5.1 Cost Segmenting and Cost Structure

Base of the cost assessment is DIN 276 (Version November 2006). The essential task of the DIN 276 is an obligatory structure of the determined costs for all buildings, which makes a reliable summary of the planned investment and/or to be documented costs possible. It must be designed so that it provides finer subdivisions hierarchically according to the planning levels of the HOAI, which also correspond to the standard cost assessment proceedings.

In the standard, statements are made concerning cost assessment in the individual planning stages. We distinguish between:

– Budgeted costs determination for the cost specification (New way of determination in accordance with DIN 276 version November 2006)
– Cost estimate for assessment of preliminary planning
– Cost estimate for assessment of schematic design
– Estimated cost for assessment of implementation planning as well as the award decisions
– Cost observation for proof of resulting expenditure as well as, if applicable, for cost comparison and documentation purposes

Three cost structure levels are scheduled, which are indicated by three-digit ordinal numbers respectively.

100 Property: At first, the client himself must address the costs for the property, its preparation and development. The construction plot includes all costs connected with its acquisition. Aside from the purchase price and/or value, here we find such things as any and all charges, taxes, redemption and compensations fees that are required to be able to build on the property.

200 Preparation and development: Preparation includes costs for concrete services that are required to put the property into a condition ready to be built on. These costs can be considerable, especially for downtown properties. Development includes charges or construction cost subsidies for services of the community or the public authorities, such as access roads as well as the supply with water and electricity.

300 Building – building structure: Segmenting of the building structure is built up according to the element method. All elements of the construction of a building – from the foundation up to the roof – are summed up respectively in a rough outline.

These building elements can then further be subdivided into sub-elements all the way down to the individual service items in the BOQs. One can imagine the rough outline like a house:

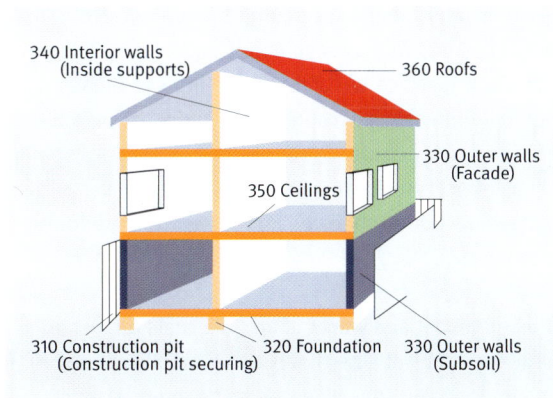

Fig. 4-53 Cost assessment via building elements

The ceilings can then be further subdivided, for example into 351/ceiling structures, 352/ceiling coatings and 353/ceiling claddings.

For a detailed cost assessment of the structural design during the more precise planning phases, the structure listed in DIN 276 does not suffice, however. It can further be equipped with full particulars by the user, whereby the essential contents are outlined verbally in the DIN.

400 Building, engineering systems: The same system applies to technical equipment, in principle, as it does for structural design. The area has, however, been assigned a cost group of its own – unlike the DIN of 1981. The "rough elements" of the engineering equipment have an advantage over structural design in that they can be directly sectioned according to services areas and/or trades, after VOB and VOL.

The DIN, as is principally the case anyway, in this area only goes up to a structure depth of three segments. Structures going beyond this have to be defined by the users themselves.

500 Outside facilities: Similarly to the building, the outside facilities also have been subdivided into corresponding cost groups. To be taken into account here is that – aside from landscape works, which are essentially grouped under 510/Property Surface Areas (Floor installation and parks) and 520/Paved Areas (Roads, paths, squares) – a number of other services need to be taken into account. This includes primarily structural design measures that, as a rule, on a larger scale belong to the shell rather than horticulture, as well as engineering equipment of a larger size. Early specification of responsibilities is already required for reasons of contract design for the planners. In these areas, experience shows there is double planning just as often as there are planning gaps whenever project management does not come up with a proper performance matrix.

600 Equipment and works of art: The costs for all equipment and works of art – that is either mobile or can be fastened without special measures be required – are summarized in this cost group.

Furniture, textiles and general devices are part of general equipment. Scientific or medical devices are described as special equipment. Other equipment includes information systems like signposts, orientation plaques, color conducting systems or advertising facilities. The costs for works of art that serve for the design of the building itself and for outdoors facilities also fall into this cost group.

700 Construction – adjacent costs: Adjacent construction costs are essentially all fees for planning and supervision services that are rendered in connection with the construction undertaking.

Costs incurred by the client organization itself, however, are also included, as long as they are in direct context to the construction work. The fee for project management, particularly, is part of it also, for this clearly falls under client performance and is not assigned to planning performance as is frequently assumed. The artistic design of components also belongs into this category. However, what does not belong here is technical production of works of art that are planned by an artist but made by third parties.

Finally, all of the costs of financing are assigned to this cost group as far as they must be included in the cost assessment as agreed. Charges like laying the foundation stone, topping-out ceremony and similar events also belong into the category of adjacent construction costs.

Project-related cost structure: The first activity is the definition of the cost assessment structure at the beginning of the project. This basic structure may not be abandoned any more during the entire project in order to guarantee cost comparison over the complete project course. It forms the "control mask" with a rough cost key, at the topmost level, for all cost extrapolations that are monitored by project management.

For large projects, the structure must be further divided, into parts of the building and individual use areas, to be able to carry out a cost assignment suitable for uses. To this end, a cost structure must be created, before the beginning of the cost assessment, in coordination with the client, which allows for aggregation on a superseding

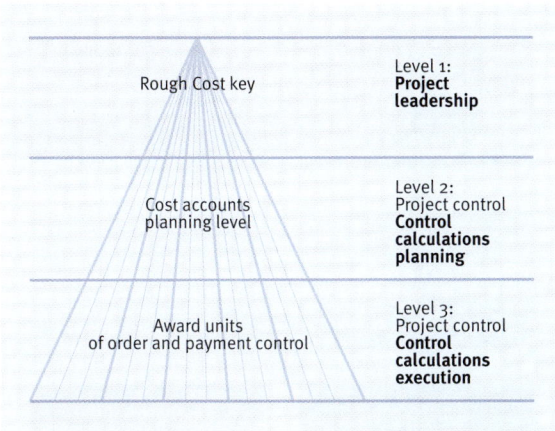

Fig. 4–54 Cost hierarchy

level. When specifying the cost structure, a sensible segmenting of the project is required. This must orientate itself on the type of building, the project boundary conditions and business management requirements.

4.5.2 Consequences of the Construction Price Trend

The costs of a building depend quite fundamentally on the time of its construction. The development of the building costs is included in the so-called building price index, which is returned by the Federal and State Statistical Offices.

The course of the construction price trend over the last four years for office buildings is represented in Fig. 4–55. It turns out that an enormous increase in construction costs was recorded particularly within the last three years. It has to be taken into account that this is a gross index and the added Value Added Tax increase for the year 2006/2007 is included.

In principle, the cost assessments need to show the cost status at the time of the investigation. This time period has to be documented. Provided that costs for the time of the cost observation are forecast, they have to be itemized separately. In principle, it has to be indicated whether the sales tax is contained or not and to which price basis the costs refer to, in order to be

able to assess the consequences from construction price increases. It is recommended to always separately itemize the sales tax.

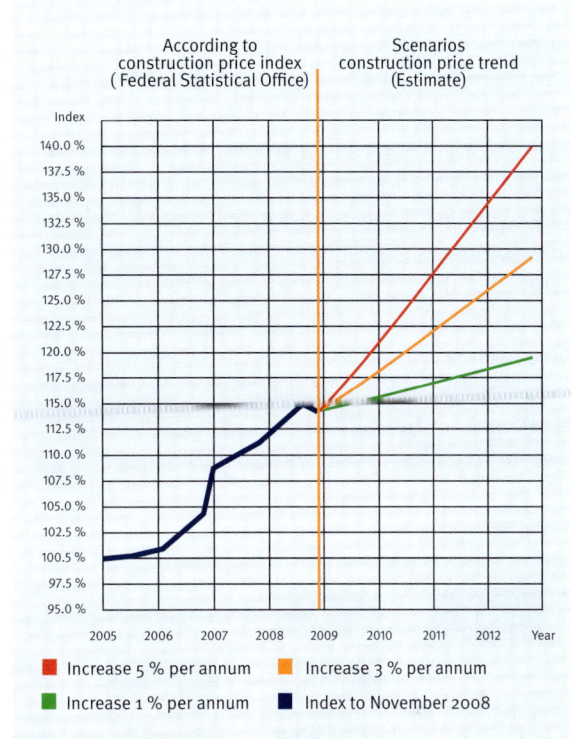

Fig. 4–55 Construction costs development and sensitivity analysis

With the assistance of a sensitivity analysis, the possible development of construction costs can be illustrated and the effect resulting from it on total project costs can be estimated. Through this, risks from price increases can be analyzed and judged – primarily for construction projects over a long period of time.

4.5.3 Cost Assessment

For every project phase, there needs to be a clearly defined cost assessment with different precision degrees, the manner and size of which orientates itself on the respective planning state. The proportion of detail increases simultaneously with advancing project progress.

Plan cost assessment and yield assessment: The decision for or against an investment is carried out during the idea phase, based on the investment cost estimate and the accompanying yield calculation.

To be able to determine a realistic return for a project, the parameters listed above must be known and the dependences taken into account. An essential role, here, is played by an estimate as exact as possible of the investment costs. For this, the author has developed the

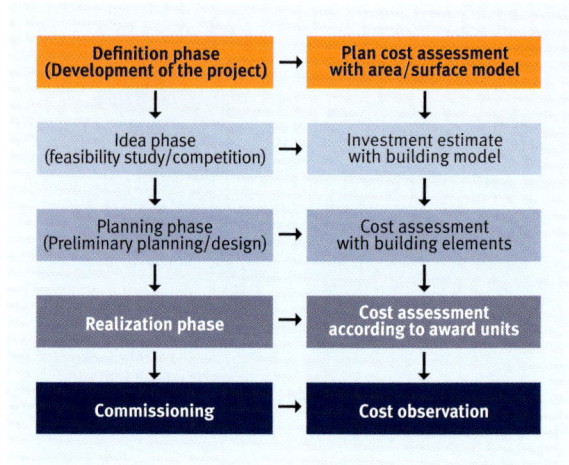

Fig. 4–56 Levels of cost assessment

so-called returns check (Fig. "Returns check" with investment estimate and yield forecast), a software that allows

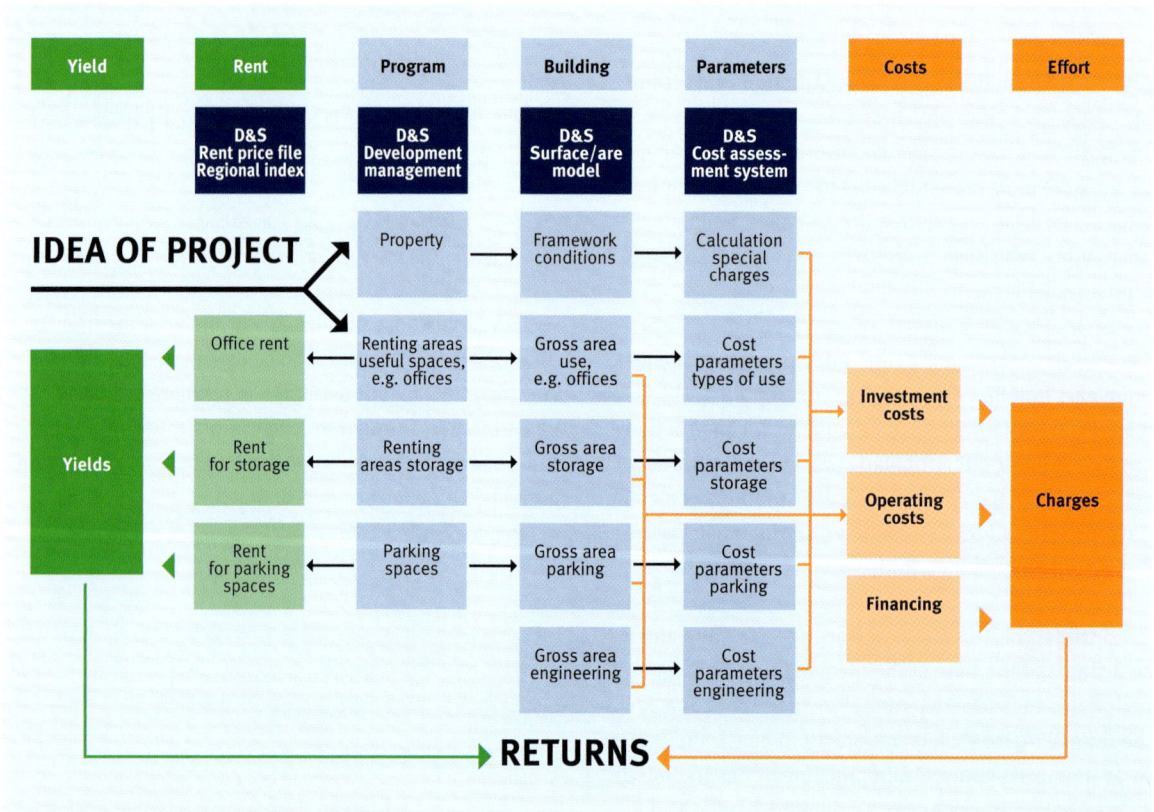

Fig. 4–57 "Returns check" with investment estimate and yield forecast

you to assess for planned projects, at quite early a time, whether a sensible return can be achieved at all.

Basis of all inquiries is the spatial program. The yields to be expected on one side and the charges to be expected on the other side are derived from it. In a first step, on the basis of requirement planning, useful space is determined and then supplemented with the necessary secondary areas (traffic, technical function, and construction area) to make a surface model.

Expected planning costs for the building are derived from the gross floor area (GFA) and use – dependent cost parameters per m². Together with the general expenses (property, development and marketing), the costs of financing and the operating costs, we thus arrive at the estimated total costs for the investor that we then compare to income estimate arrived at from looking at the available renting areas. With these data, it is possible

to assess the profitability of the investment to such an extent that a safe decision can be made concerning the further processing of the project.

Investment estimate with building model: On the basis of a surface model and by the specification of the required ceiling heights in dependence of the usage applications inside the building, the data is now transformed into a three-dimensional building model.

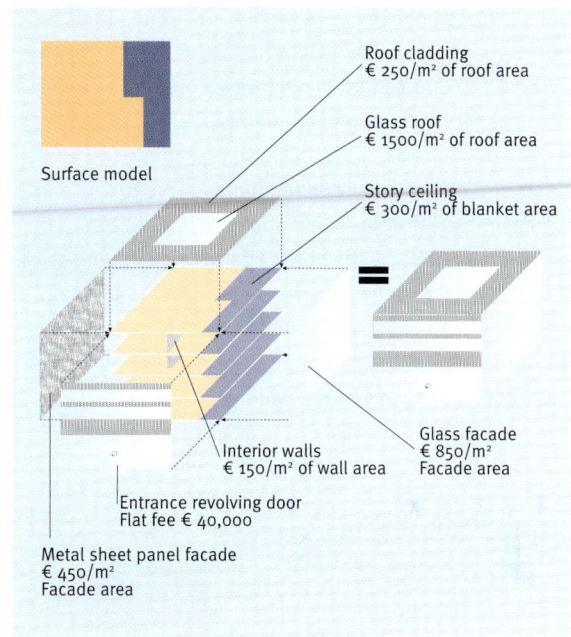

Fig. 4–59 Virtual 3 D model

Used area	Useful space		Secondary area		
	UF +	TA +	TF +	CF =	GFA
Exhibition	12.149	1.275	134	1.085	14,642
Foyer	1.160	218	14	127	1.519
Special areas	1.730	88	23	150	1.991
Catering	1.105	165	25	123	1.418
Museum shop	275	15	6	26	321
Museum education	680	60	10	64	814
Museum library	125	15	3	12	155
Museum workshop	285	23	6	27	341
Administration museum	380	114	10	50	554
Technology head offices	0	0	1.179	179	1.966
Sum total	**17.979**	**1.972**	**2.078**	**1.842**	**23.720**

UF = Useful space, TA = Traffic area,
TF = Technical function area, CF = Construction area,
GFA = Gross floor area

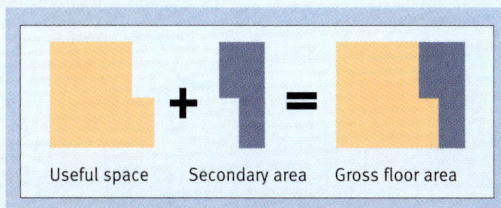

Useful space   Secondary area   Gross floor area

The required wall, facades, and roof areas can be assessed with the assistance of empirical values and the building model completed therefore. With that, all dimensions of the cost influencing rough elements required for an investment cost estimate are now available .

Fig. 4–58 Surface model

Cost assessment with construction elements and implementation types: Once preliminary planning or a building outline are available, the rough elements can now be given full particulars, all the way from construction elements to design types and leading positions.

In this, the costs, in dependence of the respective planning state, can be investigated and illustrated – by trades and at different precision steps – with the use of a suitable software (e.g. CostMonitor chapter) In the event of cost assessment by construction elements, qualities and standards of the respective construction elements need to be clarified and taken into account for the choice of the cost identification parameter.

Costs and price files of different types serve to this end, which are partly offered on the market but are also partly maintained by the market participants themselves. Generally, it has to be said that only files that are maintained by oneself allow for a truly safe statement since the attendant circumstances of the emergence of prices and cost data are of considerable importance.

Building	Type of facade	Fac. area m²	Calculated unit price DM/m²	Total price TDM
A1	Clinker facade	7.850	1.847	14.499
A2	Climate dual window	7.050	1.552	10.942
	Facade A1	**14.900**	**1.707**	**25.441**
B4 / B6	Light metal facade elements with insulation glazing	16.530	1.718	28.399
	Facade B4 / B6	**16.530**	**1.718**	**28.399**
C1	Terracotta-facade	17.570	1.218	21.400
C1	Double-skin facade	4.000	2.775	11.100
	Facade C1	**21.570**	**1.507**	**32.500**

Fig. 4–61 Determining flat fee pricing at the example of the facade

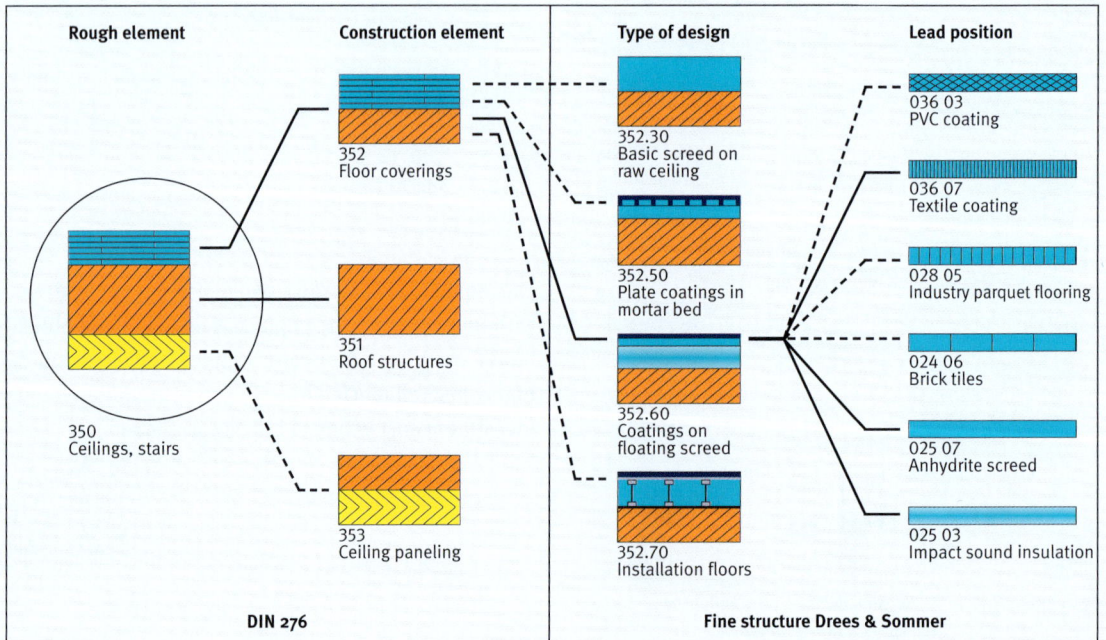

Fig. 4–60 Increasing detail in cost assessment at the example of ceiling areas (© Drees & Sommer)

Cost assessment by award units: The phase of the cost supervision starts with the beginning of the implementation planning. Target of the cost supervision is to guarantee compliance with the overall project costs by means of suitable measures.

To create the transition from element-oriented costing according to DIN 276 to trade-oriented cost supervision of the call for tender according to VOB, re-sorting the costs is required. As Fig. 4–62 shows (Transition from elements to trades),in the area of structural design, cost assessment needs to be refined down to the level of the leading positions, in order to create the transition to

performance areas and award units. An award unit is the composition of the costs of all performances and/ or services which are tendered jointly in a BOQ, such as shell construction work, paint and wallpaper work, light partition walls or suspended ceilings.

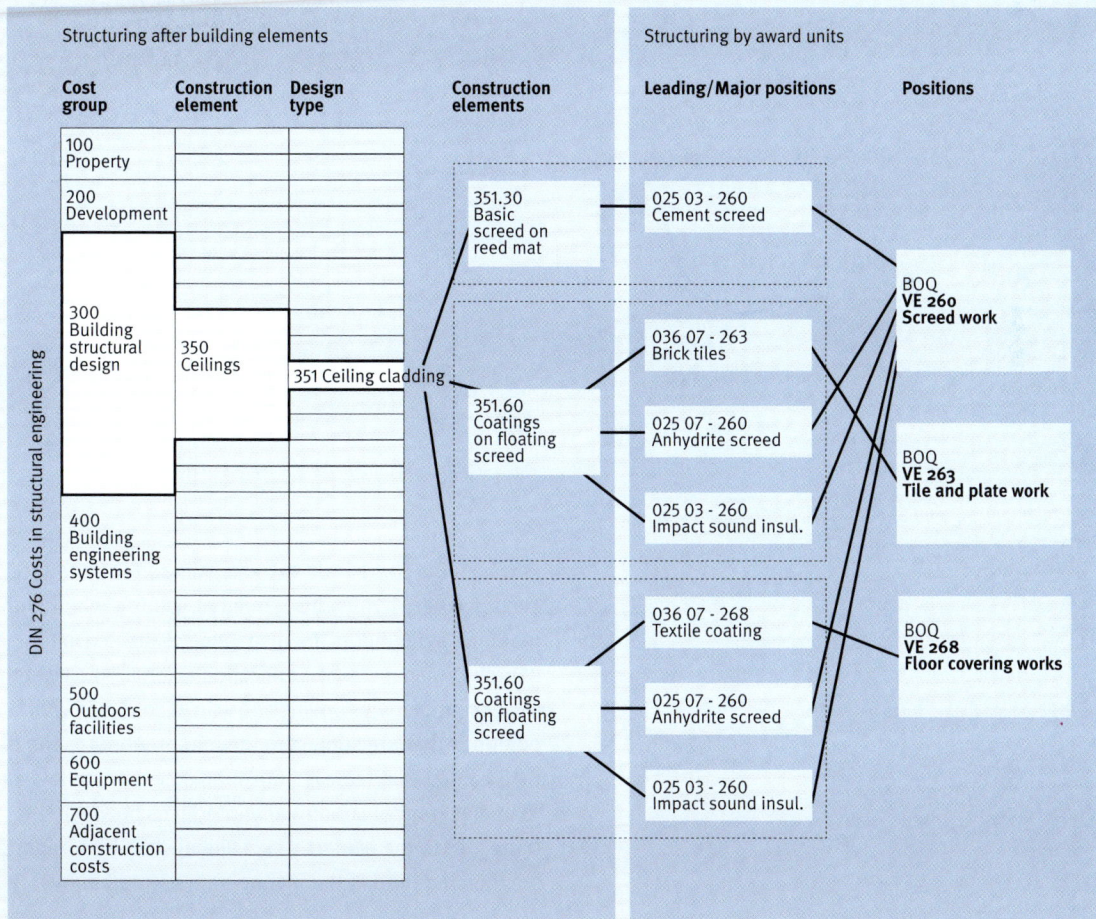

Fig. 4–62 Transition of elements to trades

4.5.4 Cost Monitoring

During the entire project execution, it must be expected that the cost bases change due to influences of every kind. To this end, all cost amendments must recorded and checked by project control via a suitable cost amendment reporting method (e.g. plan change attestations).

Cost supervision of planning: Through plan change management, changes of any kind are documented and taken to the decision stage at the appropriate client level. This means that cost coverage is already adapted constantly during the planning, which requires a close cooperation between planners and project management.

In this context, it is important to record both the cause of the cost change and the person causing it as well as the portrayal of effects on other areas and possible schedule consequences. Changes to the planned total budget are transmitted to project management as part of the report system. In case of defined, larger changes, project management must agree before the measure is initiated.

Cost supervision of the call for tender: The same as for planning applies to the translation of planning into BOQs. Review of the BOQ is the last opportunity for the project manager to actively intervene in cost control prior to the contract being awarded.

Sample project

Registration of plan amendments to design

1. Initiator

Client

2. Description of amendment(s)

The principal wishes the installation of additional glass (partition) walls to separate the occupied zone of the inhabitants from the occupied zone of the students.

3. Justification of amendment(s)

The amendment is supposed to lead to an improvement of occupancy quality and to avoid interruptions in the care area.

4. Cost implications

Supplement of the Leichtbau Company of 31.12.2005	35.000,00 EUR gross
Plus reserves 10%	3.500,00 EUR gross
Plus proportional adjacent construction costs 18 %	6.300,00 EUR gross
Overall cost	44.800,00 EUR gross

5.Schedule consequences

If a decision is made by 28.02.2005, there are going to be no schedule implications.

6. Statement project management

The above-mentioned extra costs are not part of the budget and can only be covered through a reduction of reserves.

Stuttgart, on ....   Specialist planner ....

7. Approval by client

Approved / rejected

Stuttgart, on ....   Client ....

Fig. 4–63 Example plan change attestation

Tendering control Nr. xy

Concerns:	**Paintworks**
Basis:	Bill of quantities nr. 38 As on: 08.05.19XX
Content:	The content is principally in accordance with the standard description. It is recommended to include the following additional contingency items: – Second paint coat of radiators (ca. 20 %).
Amounts:	In comparison with the cost calculation, the amounts are higher by ca. 10 %. This is justified by the uncertainty in the entrance area.
Recommendation:	The security reserve is to be reduced to 5 % since a budget reserve is already available for the entrance area. After adding the changes, the bill of quantities is ready to be sent out.

Fig. 4–64 Documentation call for tender control

At this point, there is an extremely important interface section where it must be verified absolutely whether the planning agreements are also being realized. On the part of project management, verification is required on whether the essential major positions are in accordance with the positions that are subject to the tender. Another aim of BOQ control is to remove both supplement potentials due to unclear tender content and interfaces between the trades. As to the respective individual situations, reports are then required in the manner represented above.

Cost supervision of construction: Proof of cost coverage is the central instrument in this phase. In the proof of cost coverage process, the budget of the award unit from cost planning is compared to the bid sum and the cost consequence as a clear decision base is then documented for project management. The sum of all cost recovery proofs yields the estimated cost at the time of awarding.

For comparison of bids, discounts offered must also be assessed on the part of project control in connection with advance payments through respective methods (e.g. accrued interest or discounting). If greater deviations to cost coverage proof appear, the causes are shown in order to provide project management with corresponding negotiation options. All these activities require tight coordination with the respective partial project managers for system or structural engineering as well as with the engaged architects and specialist engineers.

For preparation of the cost recovery proof it has to be taken into account that a trade-dependent budget is included in the planning for the purpose of possible supplements.

Moreover, an order-independent reserve account has to be opened for not foreseeable cost influences in order to assure cost compliance. Only overall project management may have access to this account. The method provides a cost compensation through this reserve account by covering shortages from awarding while, on the other hand, over-coverage, e.g. excess funds in hand, are paid into this account. The reserve account can be dismantled gradually over the project course, in accordance with corresponding risk considerations.

Following the awarding of the services, there is now on-going cost supervision throughout the construction process. All order-oriented data are supervised via computer. Unlike project administration, it matters particularly here that – aside from the budgets, orders and payments – risks from the construction process are also included and worked into a cost forecast at the time of invoicing.

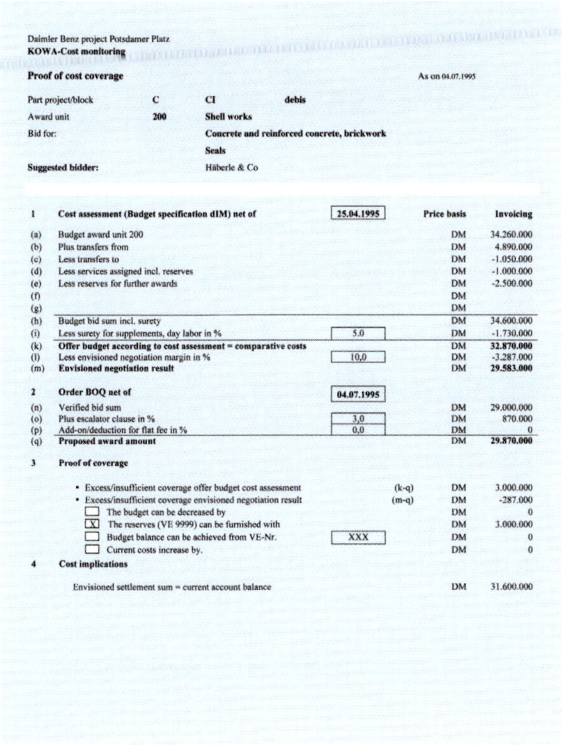

Daimler Benz project Potsdamer Platz
KOWA-Cost monitoring

Proof of cost coverage   As on 04.07.1995

Part project/block	C	CI	debis
Award unit	200	Shell works	
Bid for:		Concrete and reinforced concrete, brickwork	
		Seals	
Suggested bidder:		Häberle & Co	

				Price basis	Invoicing
1	Cost assessment (Budget specification dIM) net of	25.04.1995			
(a)	Budget award unit 200			DM	34.260.000
(b)	Plus transfers from			DM	4.890.000
(c)	Less transfers to			DM	-1.050.000
(d)	Less services assigned incl. reserves			DM	-1.000.000
(e)	Less reserves for further awards			DM	-2.500.000
(f)				DM	
(g)				DM	
(h)	Budget bid sum incl. surety			DM	34.600.000
(i)	Less surety for supplements, day labor in %	5.0		DM	-1.730.000
(k)	Offer budget according to cost assessment = comparative costs			DM	32.870.000
(l)	Less envisioned negotiation margin in %	10,0		DM	-3.287.000
(m)	Envisioned negotiation result			DM	29.583.000
2	Order BOQ net of	04.07.1995			
(n)	Verified bid sum			DM	29.000.000
(o)	Plus escalator clause in %	3,0		DM	870.000
(p)	Add-on/deduction for flat fee in %	0,0		DM	0
(q)	Proposed award amount			DM	29.870.000
3	Proof of coverage				
	• Excess/insufficient coverage offer budget cost assessment		(k-q)	DM	3.000.000
	• Excess/insufficient coverage envisioned negotiation result		(m-q)	DM	-287.000
	☐ The budget can be decreased by			DM	0
	☒ The reserves (VE 9999) can be furnished with			DM	3.000.000
	☐ Budget balance can be achieved from VE-Nr.	XXX		DM	0
	☐ Current costs increase by.			DM	0
4	Cost implications				
	Envisioned settlement sum = current account balance			DM	31.600.000

Fig. 4–65 Cost coverage proof

4.5.5 Cost Management Tools

Solid cost management is no longer realistic nowadays without the application of professional software tools. There is a wide range of standard offers on the market. This standard software is overtaxed, however, when it comes to large, complex projects and whenever specific requirements of the client must be considered also. Over the years, specific cost management tools were developed for this reason, especially for large and complex projects, which react to these enormous requirements on cost management and thus support the overall process.

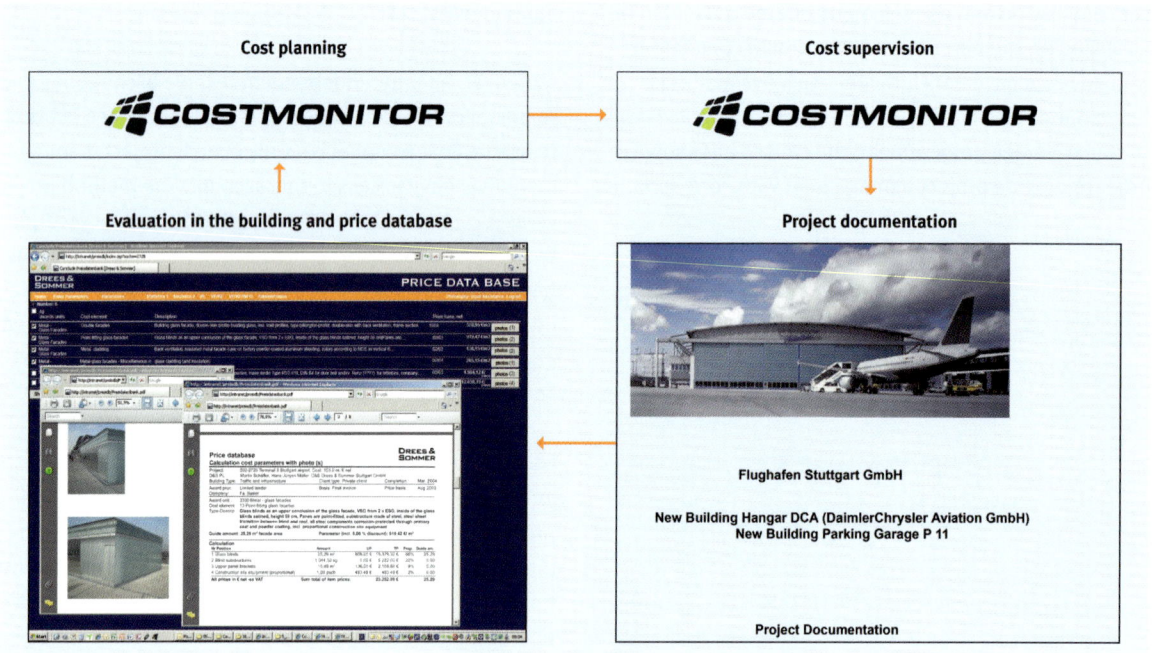

Fig. 4–66 Systematic arrangement of a professional cost management tool, example CostMonitor (© Drees & Sommer)

General operation: This is a general cost management from budget formation to cost assessment up to the end invoice and the cost determination. Further, additional features support other management services like funds drain planning or guarantee administration. Due to the comprehensive structuring possibilities, the results and cost forecasts can be subdivided and summarized on different aggregation levels at any time of the project. The budget formation is carried out according to the cost element method where all services are registered with quantity and flat-fee price. The budget formation, that is the definition of the target costs, is undertaken by assignment to components and award units. According to this project structure, cost tracking and forecast calculations are executed in the further project course through confrontation of the award-related actual costs. If budget changes arise due to approved planning changes, these are documented via plan amendment attestations.

Forecast list with orders: The actual costs contain the main order sums including the reserve budget, approved by project management, for supplements as well as payments approved for this particular order. The reserve budget is then compared with the supplement agreements, e.g. the verified and contractually agreed supplements, as well as the known contract risks. If the contract-related comparisons between budget, order and payment come

Sample project D&S

1.1 Overview list (with orders)

Sorting: Component (C) and PU

Options: incl. preliminary budgeting/orders, with discount

DREES & SOMMER

All amounts in EUR (net)

Responsible — Klaus Schwind
As on — 26.02.2009

C / PU	Current budget	Available means	Budget (covered)	Award budget current	HA+NV	Risk (Real case)	Reserve (Real case)	Pecuniary claim	S	A	Additional costs (Real case)	Prognosis (Real case)
C 01 - Office building												
Σ PU 2000 - Preparing, developing	4.587,00	4.587,00	0,00	0,00	0,00	0,00	0,00	0,00			0,00	4.587,00
Σ PU 2200 - Public access	7.645,00	7.645,00	0,00	0,00	0,00	0,00	0,00	0,00			0,00	7.645,00
Σ PU 3100 - Construction pit	63.145,05	63.145,05	0,00	0,00	0,00	0,00	0,00	0,00			0,00	63.145,05
Σ PU 3120 - Lining	0,00	0,00	0,00	0,00	0,00	0,00	0,00	0,00			0,00	0,00
Σ PU 3130 - Water lowering	1.469,40	1.469,40	0,00	0,00	0,00	0,00	0,00	0,00			0,00	1.469,40
300/001 Topbau GmbH Shellwork			682.000,00	626.500,00	15.000,00	9.500,00	365.464,67	0,00				682.000,00
Proven services		0,00	31.000,00	2.694,99	0,00	28.305,01	2.694,99	0,00				
Available means	0,00							0,00				
Σ PU 3200 - Shellwork	682.000,00	0,00	682.000,00	682.000,00	629.194,99	15.000,00	37.805,01	368.159,66			0,00	682.000,00
301/001 Glass construction facade works		320.126,00	320.126,00	302.134,92	7.500,00	10.491,08	120.000,00	0,00				320.126,00
Available means	0,00							0,00				
Σ PU 3300 - Facade, roof	320.126,00	0,00	320.126,00	320.126,00	302.134,92	7.500,00	10.491,08	120.000,00			0,00	320.126,00
Σ PU 3310 - Covering and sealing of roof	63.770,00	63.770,00	0,00	0,00	0,00	0,00	0,00	0,00			0,00	63.770,00
Σ PU 3330 - Metal-Glass facade	230.175,00	230.175,00	0,00	0,00	0,00	0,00	0,00	0,00			0,00	230.175,00
Σ PU 3340 - ... and glare protection, obscuration	0,00	0,00	0,00	0,00	0,00	0,00	0,00	0,00			0,00	0,00
Σ PU 3350 - Metal works, locksmith	23.990,00	23.990,00	0,00	0,00	0,00	0,00	0,00	0,00			0,00	23.990,00
Σ PU 3360 - Gate facilities	0,00	0,00	0,00	0,00	0,00	0,00	0,00	0,00			0,00	0,00
Σ PU 3410 - Installation floors	28.675,00	28.675,00	0,00	0,00	0,00	0,00	0,00	0,00			0,00	28.675,00
Σ PU 3420 - Screed	32.927,00	32.927,00	0,00	0,00	0,00	0,00	0,00	0,00			0,00	32.927,00

Fig. 4–67 Prognosis/forecast list with orders

up with an excess, this excess is itemized per award unit, per component and finally added up for the overall project. The budget plus additional costs is then the actual forecast, which can be determined at any time, independent of current project status.

Proof of cost coverage: Special significance when setting up an order is placed on the cost coverage proof. With the assistance of the cost coverage proof, the connection is established between the budget, e.g. the target costs, and the order, which constitutes the actual costs. To cover the order sum and the order-related reserve for supplements, the cost elements defined for this order via cost calculation are compared to the former. In the best-case scenario, this proof is already furnished to the client prior to the call for tender in order to show the maximum possible contract sum as a negotiation target with the envisioned contractor. Excess budget can be used for project reserves in order to cover for award units that are financially not quite met.

Supplement agreements: If, over the course of the project, subsequent assignments to the contractors become necessary, these are assigned to the respective main contract as supplementary agreements. Contract-related costs risks are already registered at the point of them becoming known, itemizing the full range, from of the worst-case scenario for the client via the most likely case envisioned by project control (real case) and all the way up to the best-case scenario. This means that, on one hand, corresponding counter-measures can still be undertaken in time and, on the other, early creation of a current cost forecast is possible.

Order and payment control: The order sums are contrasted with the invoices received and verified. If the payment sums plus reserves result in order excess, then these costs are put down as additional costs and become part of the prognosis. From correct logging from a tax point of view, proper handling of any invoice deductions is important. Here, consideration needs to be

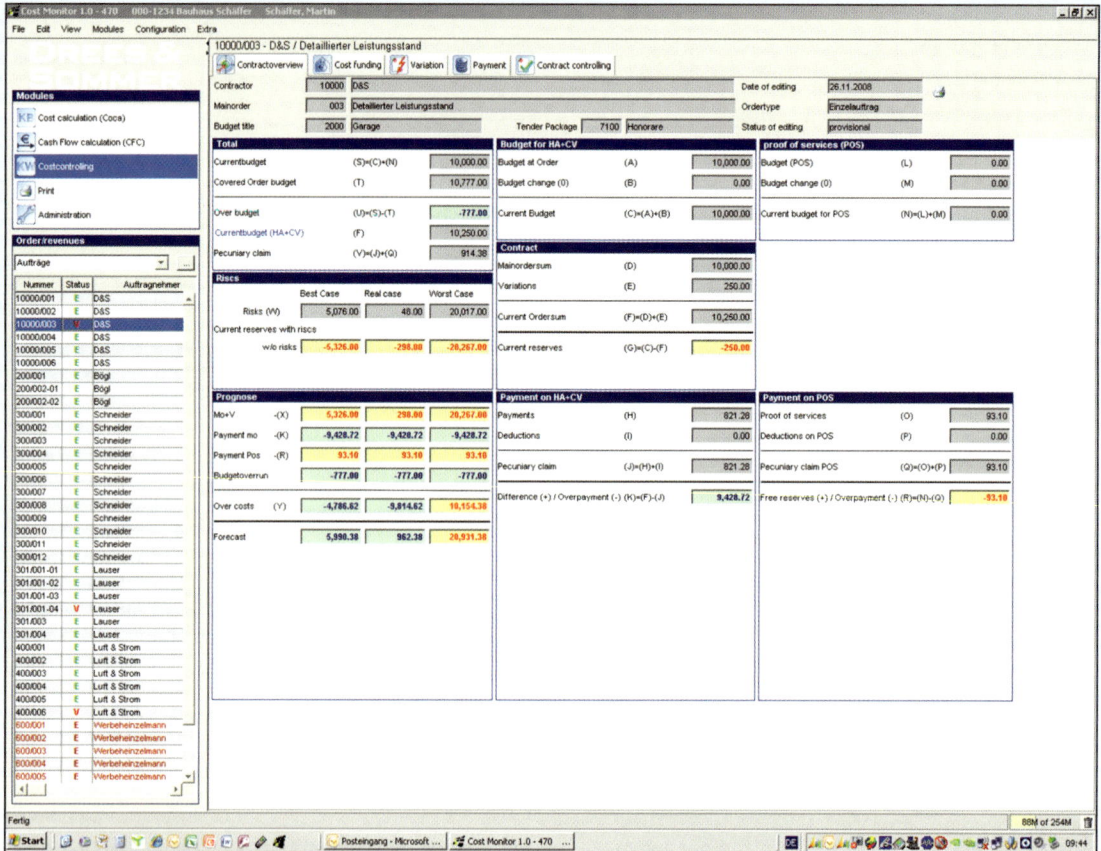

Fig. 4–68 Order overview – CostMonitor

placed on whether these deductions are so-called fee reductions, since these would lower the amount of VAT that needs to be paid.

Invoice deductions as well as contractually agreed reserves or levies as well as additional expenses through counterclaims are properly offset against each invoice. Naturally, a distinction is made between the invoices and/or manners of payment customary for construction: advance payment, payment in installments, partial invoice and final invoice as well as payout of a reserve. Fast payment reductions and monitoring of the withholding tax for construction works complete the options of invoice deductions.

Performance status control: For very critical contacts (e.g. high order volume, complex services, bad planning, "supplement-friendliness" of the contractor), monitoring of the order via pure comparison of order sum to performance status is not sufficient. Here, in addition to the relevant object monitoring, there must be monitoring on the basis of title sums. If this does not happen, a transgression in the title excavation works, for instance, would only be noticed during shell works if, at the time of the reinforced concrete work, the complete order sum is exceeded through the reported status of performance. The CostMonitor offers the possibility here of carrying out detailed performance status control for individual single orders at BOQ title level.

Cost forecasts: All relevant information of an order is summarized clearly on one page. Whether or not the order most likely will be carried out within the budget can be recognized quickly by using a red-green visualization. The data can be arranged arbitrarily and distributed in different formats via a print generator. Besides a variety of predefined lists, CostMonitor offers the possibility of defining lists of one's own with filter, sortings and groupings.

Management information: It is a good idea to file and manage appropriately, in the context of cost management, sensible information such as:

- Guarantee administration: From the time of the request of a guarantee with a corresponding note at the time of audit up to the verification of the prerequisites for the return of a guarantee
- Calculation-relevant data like discounts or surcharges
- Indications for discounts
- Calculation salary, details on surcharge rates for salaries, materials, equipment and subsequent contractors
- Contractually specified contract deadlines
- Obstruction indications or doubt registrations

Altogether, it is decisive that a universal system is available, which provides all the relevant data without effort when it is operated professionally.

Project documentation: With project documentation, completed projects are evaluated with regard to areas, costs and deadlines and the parameters documented as a basis for further projects.

Building and price database: In the building database, the data from the project documentations are recorded, processed and made available via the intranet. This includes the cost identification values of completed projects. These serve as a basis for plausibility checks for subsequent projects. The price database is an evaluation tool in which parameters can be stored centrally and then accessed via the intranet. The parameters refer to the so-called BOQ major positions or items (e.g. €/m ² floor covering) of settled projects and therefore serve as a secured and current basis for the creation of cost evaluations.

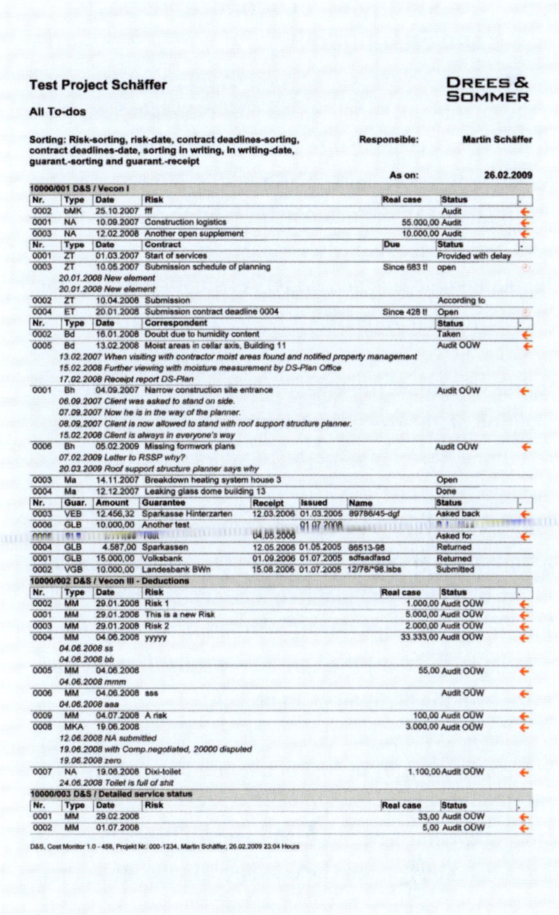

Test Project Schäffer DREES & SOMMER

All To-dos

Sorting: Risk-sorting, risk-date, contract deadlines-sorting, contract deadlines-date, sorting in writing, in writing-date, guarant.-sorting and guarant.-receipt **Responsible:** **Martin Schäffer**

As on: **26.02.2009**

10000/001 D&S / Vecon I

Nr.	Type	Date	Risk	Real case	Status
0002	bMK	25.10.2007	fff		Audit
0001	NA	10.09.2007	Construction logistics	55.000,00	Audit
0003	NA	12.02.2008	Another open supplement	10.000,00	Audit

Nr.	Type	Date	Contract	Due	Status
0001	ZT	01.03.2007	Start of services		Provided with delay
0003	ZT	10.05.2007	Submission schedule of planning *20.01.2008 New element* *20.01.2008 New element*	Since 683 !!	open
0002	ZT	10.04.2008	Submission		According to
0004	ET	20.01.2008	Submission contract deadline 0004	Since 428 !!	Open

Nr.	Type	Date	Correspondent	Status
0002	Bd	16.01.2008	Doubt due to humidity content	Taken
0005	Bd	13.02.2008	Moist areas in cellar axis, Building 11 *13.02.2007 When visiting with contractor moist areas found and notified property management* *15.02.2008 Further viewing with moisture measurement by DS-Plan Office* *17.02.2008 Receipt report DS-Plan*	Audit OÜW
0001	Bh	04.09.2007	Narrow construction site entrance *06.09.2007 Client was asked to stand on side.* *07.09.2007 Now he is in the way of the planner.* *08.09.2007 Client is now allowed to stand with roof support structure planner.* *15.02.2008 Client is always in everyone's way*	Audit OÜW
0006	Bh	05.02.2009	Missing formwork plans *07.02.2009 Letter to RSSP why?* *20.03.2009 Roof support structure planner says why*	Audit OÜW
0003	Ma	14.11.2007	Breakdown heating system house 3	Open
0004	Ma	12.12.2007	Leaking glass dome building 13	Done

Nr.	Guar.	Amount	Guarantee	Receipt	Issued	Name	Status
0003	VEB	12.456,32	Sparkasse Hinterzarten	12.03.2006	01.03.2005	89786/45-dgf	Asked back
0006	GLB	10.000,00	Another test		01.07.2008		[illegible]
[illegible]	[illegible]	[illegible]	[illegible]	04.05.2006			Asked for
0004	GLB	4.587,00	Sparkassen	12.05.2006	01.05.2005	86513-98	Returned
0001	GLB	15.000,00	Volksbank	01.09.2006	01.07.2005	sdfsadfasd	Returned
0002	VGB	10.000,00	Landesbank BWn	15.08.2006	01.07.2005	12/78/*98.lsbs	Submitted

10000/002 D&S / Vecon III - Deductions

Nr.	Type	Date	Risk	Real case	Status
0002	MM	29.01.2008	Risk 1	1.000,00	Audit OÜW
0001	MM	29.01.2008	This is a new Risk	5.000,00	Audit OÜW
0003	MM	29.01.2008	Risk 2	2.000,00	Audit OÜW
0004	MM	04.06.2008	yyyyy *04.06.2008 ss* *04.06.2008 bb*	33.333,00	Audit OÜW
0005	MM	04.06.2008	*04.06.2008 mmm*	55,00	Audit OÜW
0006	MM	04.06.2008	sss *04.06.2008 aaa*		Audit OÜW
0009	MM	04.07.2008	A risk	100,00	Audit OÜW
0008	MKA	19.06.2008	*12.06.2008 NA submitted* *19.06.2008 with Comp.negotiated, 20000 disputed* *19.06.2008 zero*	3.000,00	Audit OÜW
0007	NA	19.06.2008	Dixi-toilet *24.06.2008 Toilet is full of shit*	1.100,00	Audit OÜW

10000/003 D&S / Detailed service status

Nr.	Type	Date	Risk	Real case	Status
0001	MM	29.02.2008		33,00	Audit OÜW
0002	MM	01.07.2008		5,00	Audit OÜW

D&S, Cost Monitor 1.0 - 458, Projekt Nr. 000-1234, Martin Schäffer, 26.02.2009 23:04 Hours

Fig. 4–69 To do list construction supervision

4.6 Consulting on economical Aspects and Sustainability

In the chapter on construction cost management, it was shown by which measures and methods the specified investment costs could be adhered to. However, it is an essential task of project management to reduce investment and subsequent costs as far as possible and optimize the planning to this end under constant consideration of the intended purpose.

At first, of course, the investor is interested in the optimization of the investment costs, which essentially determine the amount of the required rent. The project manager, as a rule, works here against predefined budgeted costs, which only can be achieved by professional optimization of the essential cost factors of influence during planning. The maintenance costs, which – as a rule – one dedicates too little attention to, must also be covered by the rent. They are a very considerable cost factor over the lifespan of an object.

The essential types of cost of the building operation are the cleaning costs and the energy costs. Depending on conception, more or less high costs for flexibility and/or organizational changes by the users finally arise, up to the third-party use capacity for the investor. And, last but not least, demolition has already to be taken into account for the new building since construction is already the stage when the subsequent disposal costs can be influenced strongly.

Most of these types of cost are taken into account at the awarding of the DGNB certificate for sustainable design. Planning-independent costs like administration, insurance, garbage disposal, general services etc., are not part of the optimization task of project management.

4.6.1 Investment Costs

At first the erected construction volume influences the investment. So, for example, different organization forms in office building construction lead to very different ground plan types due to the different building depths. With an identical gross floor area, this has a massive effect on the amount of the – expensive – facade areas that need to be produced, as is shown in Fig. 4–71. A cell office in a comb structure requires almost double the facade area of an open plan structure. On the other hand, combination offices need only insignificantly more facade area than an open plan structure, provided there is intelligent ground plan design.

		Types of cost					
		Investment	Service and maintenance	Cleaning	Energy, Green Building	Flexibility of the organization	Demolition and recycling suitability
Cost influence parameters	Function/organization	○		○	●	●	
	Profitability of corp. space NF/BGF	●		●		●	
	Compactness envelope/BRI	●	●	●	●	●	
	Construction shell	●			○	○	●
	Construction facade	●	●	●	●	○	●
	Construction roof	●	●		○		●
	Construction development	●	○	○		○	●
	Engineering concept	●			●	●	●

Fig. 4–70 Impact of the individual cost influence parameters on the types of cost

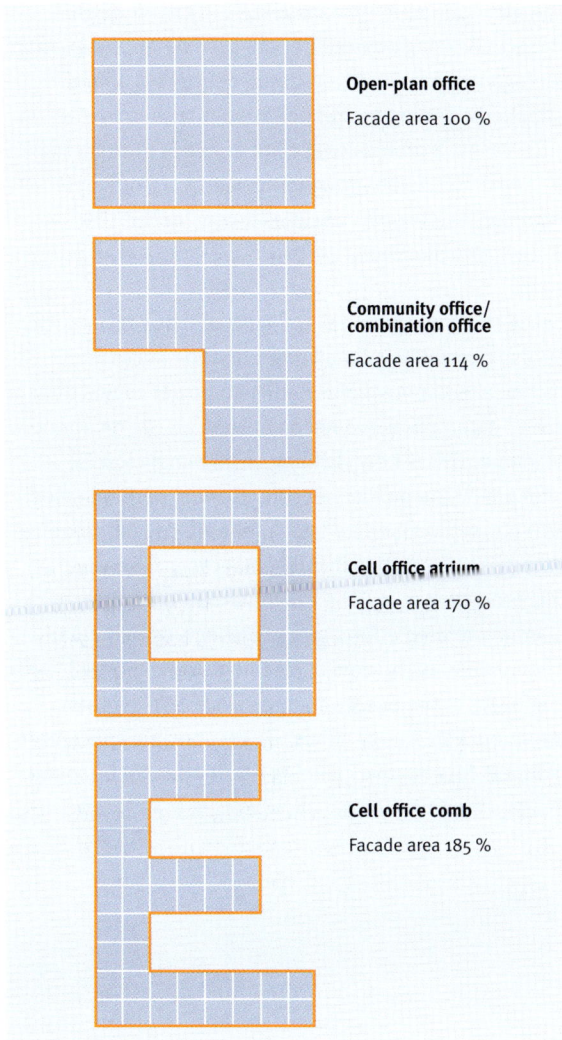

Open-plan office

Facade area 100 %

Community office/ combination office

Facade area 114 %

Cell office atrium

Facade area 170 %

Cell office comb

Facade area 185 %

Fig. 4–71 Influence of office shape on facade area

Another effective approach for the improvement of economical aspects is in persistently analyzing the room arrangement. Are all room requirements really necessary? Do all employees need a working area of their own? How many square meters are required and are sufficient for a given working area? Number of conference rooms? Canteen with a kitchen of its own, or catering? As a rule, the required useful space can be reduced considerably for the same number of working areas.

An intelligent arrangement and combination of the functions of the room program leads to a reduction of circulation areas and with that to an improvement in profitability of corporate space. This means that for each square meter of useful space (US) less gross floor area (GFA) needs to be built than with less skillful planning. Altogether, plenty of gross area can be saved by such measures, which lead to a considerable reduction of the overall investment costs.

Further savings are possible by a compact construction style where, however, architectural interests must not be left out.

Finally, further efficiency improvements are possible through structure and material choice for the essential building elements.

This concerns primarily:

– **Floor ceilings, shell:** The required supporting distances and openings are decisive factors here primarily, which in turn are dependent on the chosen building grid.
– **Facades:** The costs for facades correspond to architectural demand, material choice, building physics requirements, protection against the sun as well as the knowledge of the facade planner regarding detailed planning. Elemental metal facades are considerably more expensive than perforated facades with plastic windows.
– **Roof:** As a rule, very few saving potentials are available for the roofs if they have been carefully designed.
– **Development:** The effort required for floor, wall and ceiling is determined on one hand by building acoustics and sound engineering requirements and on the other hand by the qualitative and architectural demands.

A further essential cost factor, finally, is the complete engineering system for room conditioning, lighting and power supply.

4.6.2 Subsequent Costs

Service and maintenance costs: The later costs for maintenance are quite fundamentally influenced by the quality of the planning and implementation.

The more, at the stage of detail planning, emphasis is placed on long-term quality instead of the cheapest possible design, the lower the maintenance costs will be later on. In this place a clear conflict of interest kicks in between the building creator – whether this is the project developer or a general contractor – and the investor. The investor must meet maintenance costs from the rent and therefore is interested in quality as good as possible. At the same time, however, he or she would like to pay as little as he/she can get away with, which means products and details as cheap as possible for the creator. The task befits the project manager as a person representing the interests of the investor, in this place, to guarantee a quality as high as possible at economically acceptable costs.

Cleaning costs: There are enormous reduction potentials if architects and clients include the later cleaning in their plans already early on. What many do not know: The cleaning costs amount for an office building to between € 0.20 to € 1.50 per m² of cleaning area with clear upward leeway. Costs of 20 to 40 per cent could be saved if one would already take into account the later cleaning

charges during preliminary design. In extreme cases, the economizing possibilities are even better: for example for facades that can be accessed neither via a firmly attached cleaning basket nor through a cherry picker. In such cases, a professional climber needs to clean the facade under acrobatic conditions and, according to Sommer, needs up to 20 times longer for this than with a moveable platform.

Easily cleanable material and surfaces, as well as good access possibilities for all areas and the necessary technical equipment plus sufficient power connections are all criteria that one needs to keep an eye on during planning. This is how cleaning costs remain low for later on. "If the project-specific cleaning performances differ from the standard tasks, then we consult cleaning enterprises for advice in individual cases." However, this procedure is still the exception. Unfortunately, planning-related efforts are, at present, usually clearly undervalued in the area of cleaning. While the civil engineers or the energy managers are aligning their attention with energy efficiency, responsible specialized planners for effective cleaning usually cannot be found inside a planning team.

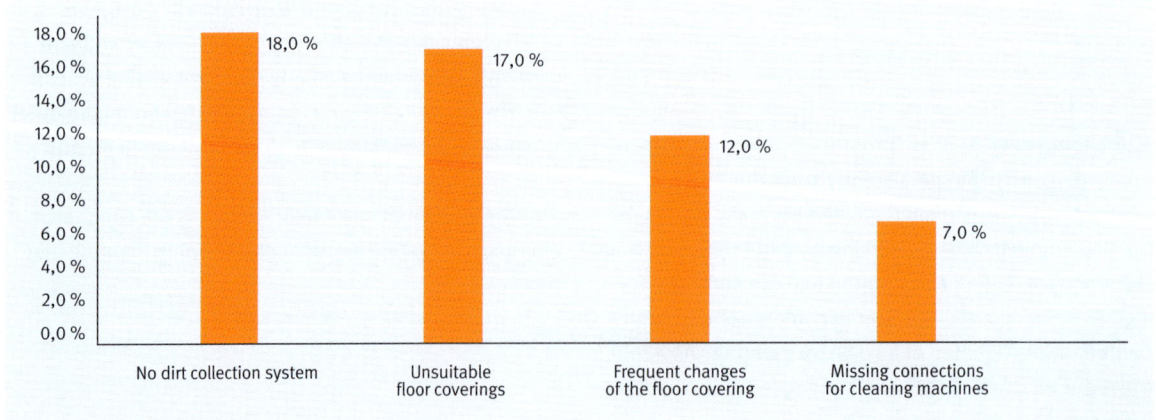

Fig. 4–72 The four main causes of high cleaning costs

Energy costs – Green Building: To achieve energy costs as low as possible for later operation, project management must completely clarify the requirements on the energy system right at the very start of the planning conception (see Fig. 4–73).

It is decisive to define whether one wants to use alternative energy, such as geothermal energy, since this is possible only with low temperature heating. In turn this would require surface-oriented heat distribution, like by concrete core activation, which in turn needs to be defined for preliminary planning at the latest. The interaction with building ventilation and lighting as well as with useful current also needs to be clarified.

The required energy demand should be reduced to the economically sensible minimum, in extreme cases in residential construction even down to zero. In any case the specifications of the Energy Savings Ordinance (EnEV) have to be adhered to. With intelligent planning, the specifications of the EnEV can be implemented in a manner so that, in the end, the saved energy costs largely compensate the increased capital charges for the required measures.

The reduction of heat and cooling energy requirements through structural measures is essentially achieved via the building envelope and the storage ability of the structure itself.

The building envelope consists of three areas:

– Cellar ceiling
– Facade (Walls, windows, doors)
– Roof

The most effortful area and the most demanding one from a planning point-of-view is the facade, which as a rule presents the biggest contact area with the ambient air at the same time. Weak points are primarily windows, outer doors and ceiling connections (heat bridges). The outer walls can constructively be designed with insulating materials or wrapped in insulation material. The windows must be designed at least in double-insulation-glazing version but better still, with triple glazing. The same applies to the insulation of the outer doors

Heat insulation of cellar ceiling and roof is comparatively easy to manage if technically impeccable details are planned.

To obtain the optimal result for investor and users, several of the overall alternatives ought to be simulated and the respective total energy requirement as well as the corresponding primary energy requirement determined. The respective investment and energy costs must be determined and compared on the basis of the different energy requirements of the variants.

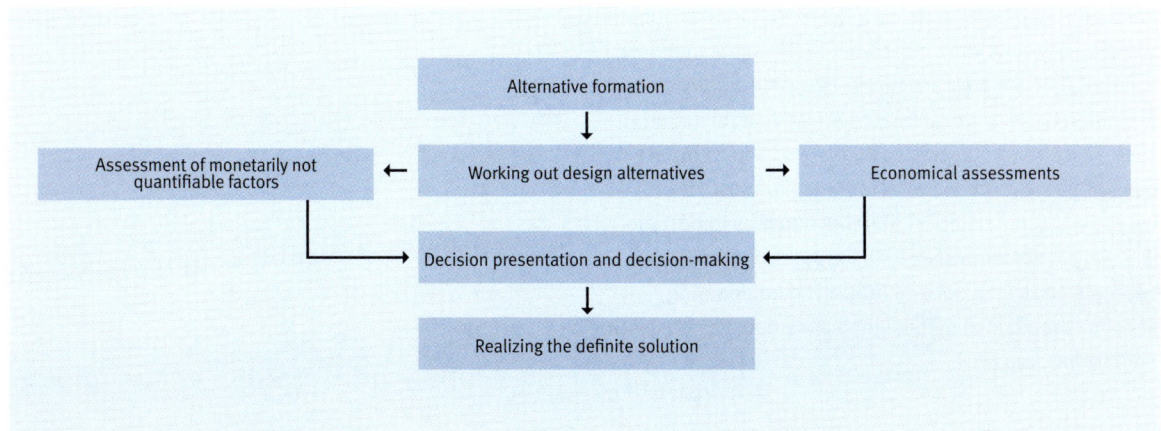

Fig. 4–73 Process: energy optimization

At the same time, the monetarily not quantifiable factors must be assessed, for instance according to the value analysis method. This includes design and function as well as technical and temporal feasibility. But coziness and comfort must also be adequately taken into account.

The variants must be processed ripe for judgment so that schematic design can be entered into with a well-secured solution. The decision should primarily also take into account sustainable criteria besides the economic ones.

4.6.3 Organization and Flexibility

There are organization changes on the agenda already before and constantly after moving into the building. Employees move, corporation sections are shifted. In these cases, it shows how flexibly planned the ground plans and the development really are. In office building construction, for instance, combination offices are clearly at an advantage since employees simply move from room to room and no partition walls must be moved. For cell or cubicle offices – primarily in the case of occupants operating at many hierarchy levels (grids!) plenty of reorganization work descends on the organizers. It has to be checked here whether, despite higher costs, investment into an easily realizable assembly wall system is not more favorable in the end than permanent demolition and new construction of walls.

For the user, attention to third-party use capacity is of great importance. What to do if the tenant moves out and the property cannot be rented out in the same manner or for the same use area any more? Planning variants where the different uses are simulated are advisable in this case. More generous area concepts and less compact envelopes are more suitable for these interests than very specific and compact buildings. Just as important is a shell in skeleton style with load-bearing walls and separate grids for shell and fitout. The engineering concept should also be designed as variably as possible and flexibly, too.

4.6.4 Demolition and Recycling Capacity

Unlike in the past, an ordered renaturation process increasingly gains more significance for cost and sustainability points of view. This means that attention has to be paid by project management already during the call for tender, as well as in detail and workshop planning, that the different materials are easy to separate and as recyclable as possible. Otherwise, considerable costs and disposal problems come to the investors in future.

4.7   Quality Management

Quality is produced, first of all, in the planning. During construction, you can only improvise and improve anymore. This means structural deficits arise at planning and then are reinforced still during implementation.

Quality surveillance, planning and call for tender: The complexity of structural and technological conditions makes it difficult for the clients and architects to be able to assess the work of the specialized planners and later also of the construction companies. So, uncertainty frequently prevails over whether the – for the operation of a building indispensable – energy concepts optimally meet their use and purpose. Modern buildings must achieve various economical and functional aims, of which some partial ones are even contradictory in themselves. As if short planning and construction times were not enough of a challenge, the buildings must also observe ecological requirements. However, the more complex the specifications are, the more system errors creep in, which means additional cost or quality loss for the client.

Quality control must start during the early planning because the course is set for the project success here. It has to be made sure that the requirements on the building are assessed in form of a careful inquest into fundamentals. Building on this, the accompanying economical and cost comparison calculations must be carried out, which also consider the later operation, maintenance and upkeep of the building. Unclear task definitions lead to uncertainties inside the planning team and to uneconomical solutions.

Planning results must be regularly assessed, accompanying the planning, and this must be done according to both technical and economical criteria. These criteria are:

– Are the planning goals agreed taken into account and accomplished?
– Is the quality right?
– Are decisions imminent that lead to black spots in the planning?
– Do the technologies complement each other? (e.g. Interaction facade, heating and cooling)
– Are the different trades coordinated with each other?
– How are the systems dimensioned? Are simultaneity factors taken into account? Are reserves superfluous?
– Are the services planned in a manufacturer-independent manner? Are there alternatives for this conception, which are less expensive?
– Are there any uncertainties?
– Is the cost estimate plausible, does it correspond to the planning and the budget?

Quality control in planning must be supported by calculations, follow-up calculations and plausibility comparisons of its own.

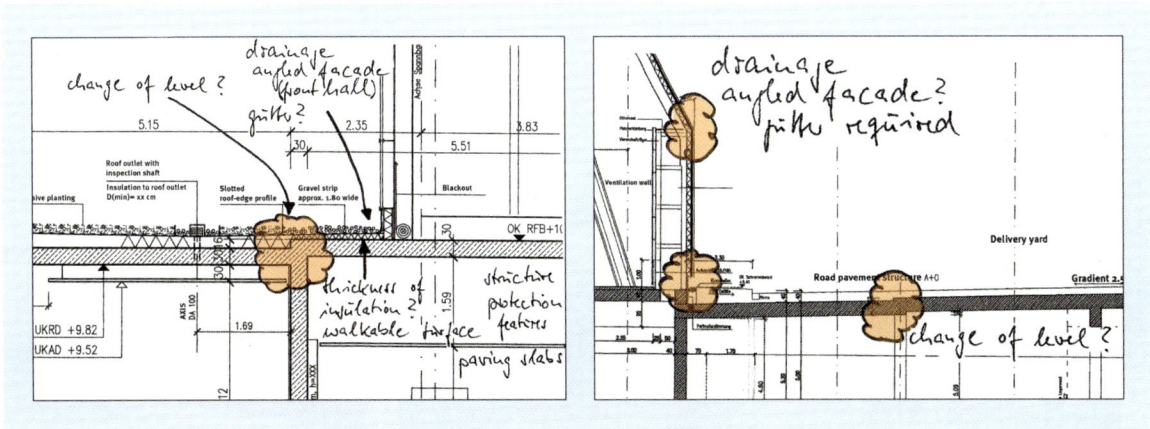

Fig. 4–74 Quality control planning

As soon as the planning phase has been completed, the services are outlined in Bills of Quantities (BOQs) with the aim of finding the suitable executing firms. The greatest deviations from the planning aims agreed on appear in the transition from planning to tender. Additional wishes are often included in the BOQs, which can lead to cost overspill. Therefore, these BOQs must be carefully reviewed. In this, the following topics are in the forefront:

– Are all services properly outlined?
– Are the services described in a manner that is manufacturer-independent?
– Does quality correspond to the agreements?
– Are extra services, preliminary notes etc. fair, practice-proven and complete?

Technical-economical controlling can also provide effective support in the area of offer evaluations: Are side offers technically and economically interesting, perhaps, does the market reflect particular signals due to a short-term shortage of goods, are there alternatives for such situations, etc.?

Quality surveillance, construction (supervision): The next quality control is required at the stage of regular building business. Here, quality must be monitored depending on construction progress so that no series faults creep in, or structural facts are created, which lead to compromises. Through regular quality control, unpleasant surprises are avoided at the time of acceptance.

The 10s rule: An experience-based rule of thumb from quality management says that troubleshooting costs increase around the factor 10 if problems are not already avoided during planning.

The 80 – 20 rule: The experience-based rule of dissimilar weighing says that 80 % of problems can be prevented with only 20 % effort. Another 10 % of the fault elimination requires, however, 80 % effort.

Technical and economic controls must, in any case, be carried out by experts of the trades. This secures the technical qualification and primarily also the acceptance of the planner by the inspectors. Furthermore the inspector should have a high level of social competence, in turn, for acceptance by the planning. The inspector should not act merely as an inspector but also as a coach, adviser and sparring partner because best results in a team can be achieved by mutual stimulation.

The results of the technical-economical controlling are transparency, guarantee of economical planning concepts and functionality. There is also decision certainty, cost safety, cost-cuttings for investments and operation as well as appointment certainty. Frequently forgotten but immensely important to the clients: the good feeling caused by the controller that everything is running correctly.

Fig. 4–75 Quality surveillance execution

Monitoring acceptance and error removal: With the ever increasing complexity of construction undertakings as well as for the ever faster construction execution with high simultaneity factors, as a rule, error routines are no longer possible. This inevitably results in a solid accumulation of defects. Substantial significance, therefore, is placed on error detection, administration and assessment during the final phase of the construction process.

Efficient and transparent control of error removal is a prerequisite to guarantee commissioning/handover as smooth as possible and agreeable to the user.

The following essential requirements on the error tool therefore arise:

- Recording of all necessary data in the process of elimination (Location, value, deadline, acceptance impeding)
- Easy sorting and filter possibilities

- Definition of rights for individual user groups (Principal (PRIN), Construction management of contractor (CMCON), Contractor (CON), Sub-contractor (SC))
- Easy accessibility for all involved

Besides the contractual prerequisites, it is mandatory to define the necessary database to be included and to anchor it in functional specifications. With the data, care has to be primarily taken that an easy as possible filtering and sorting is available later on. One distinguishes between standard entries (quantity, location, description, deadline, monetary assessment), effects on commissioning, status of the removal, serious or light malfunction or error. Moreover, it makes sense to cluster according to the major trades (Shell, roof, facade (FAC), Building Services Engineering (BSE), development etc.) and according to areas or parts of a building.

To make possible only minor defects at the time of handover to the occupant, the process needs to be structured

Major trades	Standard					Assessment and/or opening		Status malfunction removal			Quality defect/malfunction				
											Elimination required			Reduction possible	
											Consequence				
	Number	Coding for exact position definition	Description defects	Deadline	Monetary assessment 1-fold/3-fold	Deadline-relevant with respect to 6-30-7	Remark	Reported free by CON	Confirmed free by PRIN	Objection by CON	Assessed by CON as not removable (reduction)	Function-and use suitability	Maintenance/upkeep	Strong visual impairment	To be removed with 1 service/ maintenance interval	Slight visual impairment
Shell																
FAC/steel-girder construction																
Roof																
Development/ fitout																
BSE																
Outdoors areas																
Sum total																

Fig. 4–76 Example: malfunction/error clustering

from service observation (defect identification) to actual acceptance and it must be thoroughly planned down to the detail from a schedule point of view.

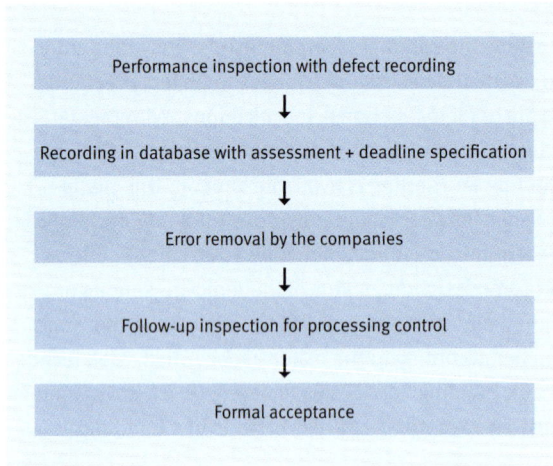

Fig. 4–77 Performance recording, error/malfunction detection

This process should be envisioned for the 6 –12th calendar week in schedule planning and, especially, also categorically implemented.

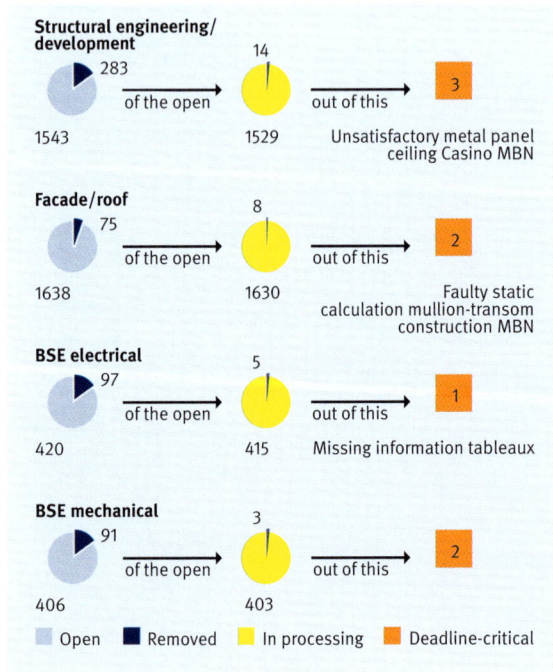

Fig. 4–78 Report structure

A simple and as thorough as possible report system is essential, besides process control of the defect identification and tracking of processing, to make an acceptance by client and occupant possible.

Those defects along the critical path need to be clearly represented, especially, that would prevent acceptance and/or commissioning and they need to be subject to a separate tracking process (Task force).

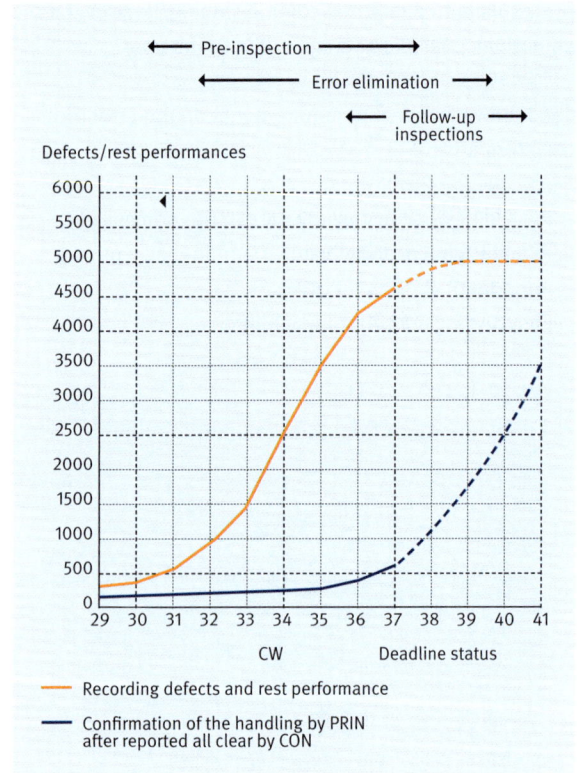

Fig. 4–79 Course of defect removal

With a stringent error management accompanying the construction process, error removal by the time of acceptance can be clearly accelerated and the irritating follow-up after occupation can be kept to a minimum.

4.8   Project Communication Management

Large or complex projects are no longer feasible nowadays without an efficient project communication system because of the number of people involved – who are usually also separated spatially. These communication systems run on central servers on which the innumerable data are exchanged, managed and archived by means of an expert software.

Web based project communication systems have gained acceptance on the market. They offer an added value for all those who are involved in a given project:

– As a complete project archive
– By high quality of the documentation
– By a high degree of availability
– Through increased safety
– By maximum transparency
– By process optimization

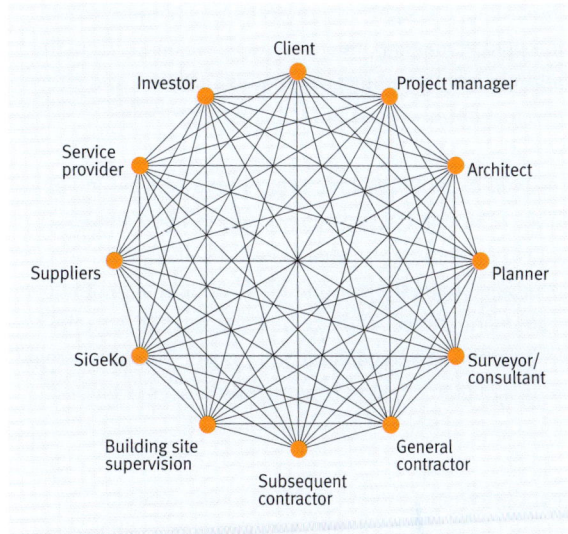

Fig. 4–80 Project Communication Management

A prerequisite for this is that the systems exactly show the agreement of the project organization and that users can handle project time according to their roles, their rights and duties and also make use of this opportunity.

INFORMATION	Information system for compressed information provision to decision makers
DOCUMENT MANAGEMENT	Central filing of all project relevant documents, plans and drawings
MESSAGE ADMINISTRATION	Automatic notification of the project participants concerning new documents, tasks, news
TASK MANAGEMENT	Tasks and schedule tracking
APPOINTMENT MANAGEMENT	Project appointment book, incl. combination with invitations and log files
PLAN MANAGEMENT	Illustration & organization of plan sequences based on target plan lists
ERROR/MALFUNCTION MANAGEMENT	Malfunction/defect identification, administration and tracking
SUPPLEMENT MANAGEMENT	Illustration of processes as a result of obstruction messages, additional costs, doubts
DATA EXCHANGE	Interfaces to other DMS and CMS systems. Email multi-upload. Data interchange with FTP servers. Module repro institution with integrated delivery note generation
PROJECT CONTROLLING	Cockpit for compressed view of delayed tasks, news or files of the project participants. Illustration of approval processes
SEARCH	Modularly arranged search engine with fast search function
MULTILINGUALISM	User-specifically adjustable user interface for different languages
PARTIAL AND MULTI-PROJECT ABILITY	Subdivision of the complete project into subprojects (components, branch offices) with an integrated rights assignment for different project teams. Uniting different single projects into one multi-project with a rights-controlled evaluation and integrated "direct data interchange".
MOBILE USABILITY	Connection of mobile terminals (Wireless handheld, mobile telephone etc.)

Fig. 4–81 Functionalities PCM

The project manager – supported by a competent data and plan manager – must set the right course, from the beginning of the project, for the organization and later also for control of the digital planning process.

The web-based PCM, which was already developed by Drees & Sommer in 1997 and has been used successfully in numerous large-scale projects since, then serves as project room, project-related document management system and project management system at the same time within a given construction project. The aim here is to design planning and construction processes more effectively and to produce transparency.

Installation of a project communication system: The classic and most frequent application for all phases of a construction project is the project room. All those involved in the project who have an active share in the planning process receive access rights corresponding to their roles: the client, his/her representative and all contractors, like planners, experts and executing companies. The use of the project room must be estab-

lished by contract and defined in a "project standard" as an addition to the organization manual.

The project room functions as a permanently up-to-date project archive and, therefore, is the ideal basis for documentation of every phase of the project, particularly however at the completion stage of a project. Prerequisite here are that all "project relevant" documents are provided. Aside from CAD models and plans, this also includes the complete correspondence, the approval documents, contract, costs and appointment data. Basically, it involves everything that is subject of project meetings or is discussed between those involved in the project.

Via the project room, the complete planning process of a 2 D and 3 D planning can be shown, something that is indispensable primarily for general planning and general construction management. If there is the requirement of a complete documentation of data, information (e.g. decision documentation) and investigating of documents, PCM also can be used as document management system (DMS).

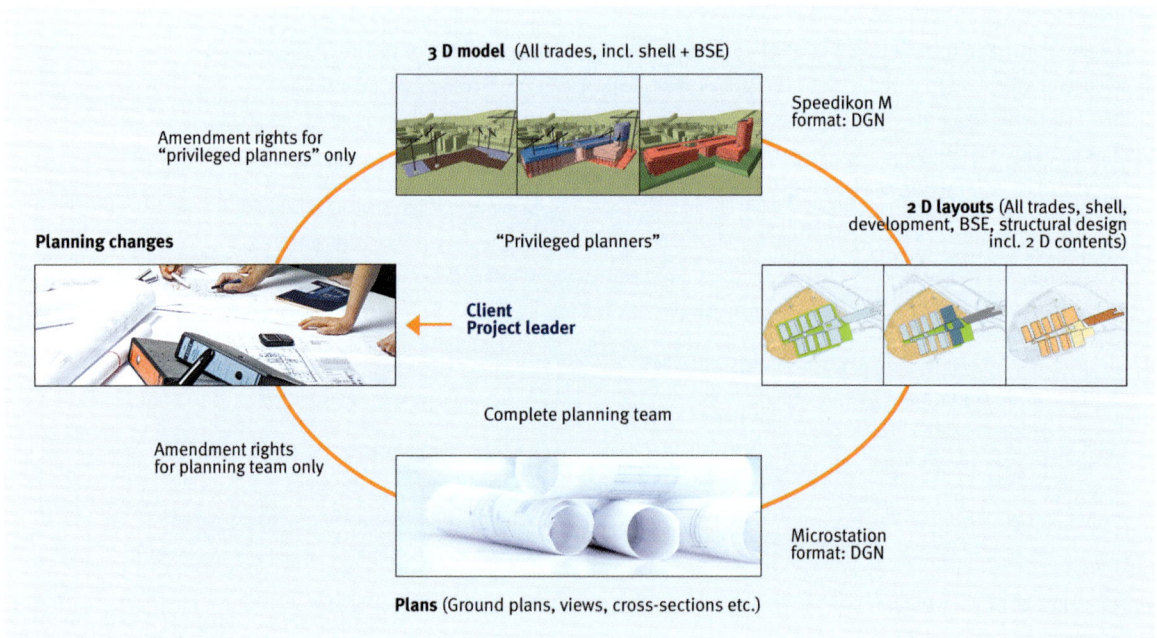

3 D model (All trades, incl. shell + BSE)

Speedikon M
format: DGN

Amendment rights for "privileged planners" only

Planning changes

"Privileged planners"

2 D layouts (All trades, shell, development, BSE, structural design incl. 2 D contents)

Client
Project leader

Complete planning team

Amendment rights for planning team only

Microstation
format: DGN

Plans (Ground plans, views, cross-sections etc.)

Fig. 4–82 Project room PCM

If large information quantities (e.g. knowledge management) are to be provided to a select circle of the people involved, PCM can be designed also as a Content management system (CMS).

The project room should represent the only communication platform for the project and therefore also document all "activities" (e.g. upload and download and/or distribution of data and information). It will then provide an important added value as a project-specific document management system.

It is necessary to advise decision makers and project leadership at the beginning of the project, to supervise all involved in the application of the project room and to check their activities if required. This requires a lot of expert competence, particularly in the case of execution of plan management tasks in the project room.

Multi-project room: The multi-project room differs from the project room by the simultaneous illustration of several projects or subprojects. It is therefore frequently used by institutions with active and continual building activity (new building, building upkeep and reorganization). The benefit for them is that the project room can be precisely configured to their requirements and considerable reduction potentials can be achieved by the frequency of the application.

Another unusual feature of the multi-project room consists of that fact that, besides the role dependent rights, project dependent rights can be shown in addition. This means that different project teams can use the platform in parallel without being able to enter into communication with others or gaining access to data of other projects. Requirements on erection, operating and reorganization processes can be optimally shown through this.

In the following section, some screenshots are shown to provide an impression of the functionality.

This allows, at any time and any place, for both information for the decision level as well as all data for the work level to be provided and for the required communication to take place.

Fig. 4–83 Representation of construction progress in the information part

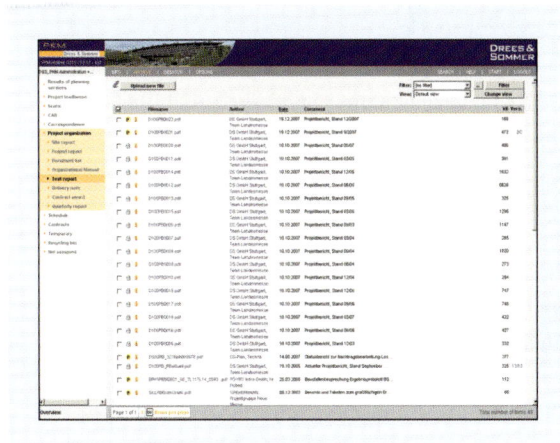

Fig. 4–84 Workflow documentation

Fig. 4–85 Planning information

4.9 User and Tenant Coordination

Unfortunately, smooth and in the long run successful completions of a contract between client/ investor and tenant are not a matter of course.

There is a whole number of causes for bad surprises:

– Tenant management is often started too late.
– Tenant often cannot formulate any really precise requirements on a building.
– Lease contracts are legal entities where the actual construction process for a building is not reflected.
– Structural, technical and organizational consequences of contractual promises are easily under-estimated. Incalculable schedule and cost problems can result.

Successful letting starts with the preparation. Therefore, tenant management must already start at the time of construction for the basic development. As long as there are still the greatest available freedoms for decision-making, tenant management already enters into agreements with the planners and specifies roles and tasks for letting and marketing. Organizational structures, which provide for solid processes and sequences, arise from that.

These organizational structures prove themselves as soon as the first tenants come to inquire. The rapidly increasing complexity of several qualification profiles, appointment wishes and know-how levels on the tenant side, as of this time, can only be efficiently controlled by a professional tenant management:

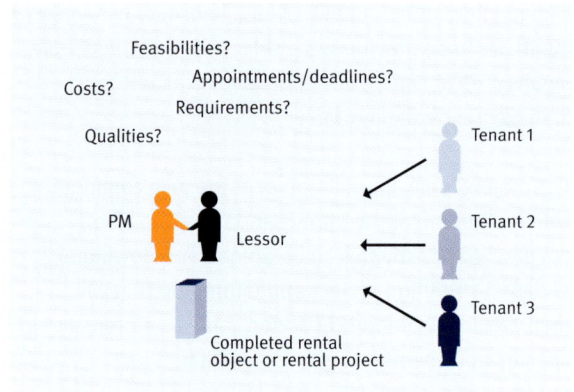

Fig. 4–87 User and tenant coordination

– Registration of the exact specification of the requirement
– Clarification of what is technically and structurally practicable
– Tenant special requests: check
– Logistical consequences: analyze
– Assess rough time frame
– Limiting costs and showing consequences for the lessor
– Create bases for deadline acceptances (incl. approval ability and building application requirements)
– Warrant secured leases as regards to content and schedule

Through good coordination and tight project management, considerably earlier dates of occupation are finally feasible than with less coordinated processes. Decisive for the returns (aside from construction of a building true to cost and schedule specifications) is that the building can be swiftly occupied by the tenants and true to requirements, exactly to the conditions agreed on by contract.

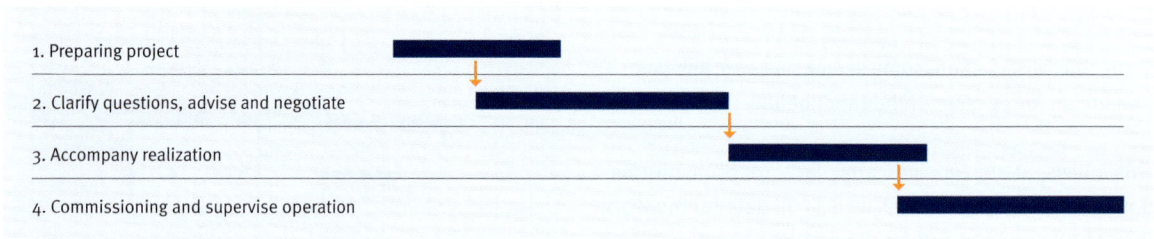

Fig. 4–86 Processes of organization for tenant management

5 Management Varieties

Generally, a client can perform all management tasks in-house. He or she can also assign them, either in total or in part, to professional project management companies. Whether he or she decides to do so, and to what extent, depends on the special situation of the client. There are numerous points of view about this and just as many theoretical explanations. What is decisive, however, is that practical experience and scientific investigations have shown that early involvement of professional, external project managers leads to an optimization potential of 10 to 15 per cent of construction costs, maybe even more. This depends on professionalism levels and competency of the external project manager as well as on the extent of the responsibility areas assigned.

This chapter outlines practical solutions in their different varieties, as they result from over 35 years of experience by the author. The respective disadvantages and advantages for the different varieties are often viewed differently by different clients, depending on their individual priorities.

5.1 An Overview of the different Varieties

In practice, the varieties here shown have proven to be the conventional ones, whereby the Construction Management varieties have arisen from the Anglo-Saxon construction field and can be applied internationally.

The varieties shown have a step-by-step built-up, whereby the client keeps assigning increasingly more responsibilities to his/her external manager right to the point of complete takeover of entire performance. Accordingly, the level of trust also needs to increase

simultaneously to the responsibilities assigned and, here, certain monitoring mechanisms are to be recommended.

Project controlling	Project control	Project management	Construction management (CM)	General construction management	Construction partner management (CM at risk)
					Building construction with fixed fee or GMP
				Logistics planning	Building site logistics
			Supervision	Site supervision	Site supervision
			Technical & economic controlling	Technical & economic controlling	Technical & economic controlling
			Value engineering, optimization	Value engineering, optimization	Value engineering, optimization
		Contract controlling	Contract management	Contract and risk management	Contract and risk management
		Award controlling	Award management	Call for tender, Award management	Call for tender, awarding
		Plan coordination	Planning management	General planning (possibly incl. architect)	General planning (incl. architect)
		Project lead function contractor	Project lead function contractor	Project lead function contractor	Project lead function contractor
		Project communication PCM	Project communication PCM	Project communication PCM	Project communication PCM
		Process optimization and simulation	Process optimization and simulation	Process optimization and simulation	Process optimization and simulation
	Project organization	Project organization	Project organization	Project organization	Project organization
	Quality controlling	Quality management	Quality management	Quality management	Quality management
Cost controlling	Cost control	Cost management	Cost management	Cost management	Cost management
Schedule controlling	Schedule control	Schedule management	Schedule management	Schedule management	Schedule management

Excluding HOAI services — Including HOAI services

Management basic services | Management additional services | Construction services | HOAI services

Fig. 5–1 Overview of the different project management varieties

5.2 Project Controlling

When project controlling is used the client himself/herself takes over the management function and intensively accompanies the project all the way. He/she obtains professional support when it comes to controlling of schedule and costs – sometimes also for quality control.

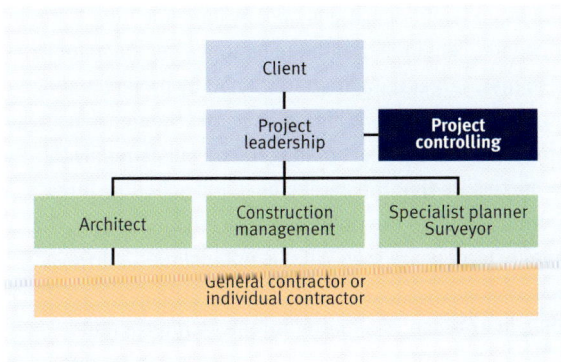

Fig. 5–2 Project controlling

Project controlling has a staff function for the project management of the client and works jointly with the same by creating control reports about actual processes in comparison to the instructions of project management.

The other project participants are in a managerial function to the client who assigns them directly. Assignment of planning is undertaken in orientation on the German HOAI, depending on whether the construction tasks were assigned or are to be assigned to a general contractor or to individual trades.

The client needs to be aware that project controlling does not execute any kind of allocative function. It merely points out recognizable deviations from the target and does not determine the target data itself.

This variety is applied by professional clients with their own know-how, e.g. with a construction department of their own, or for very simple projects. Fees are kept quite low and usually range somewhere around the area of ca. 0,7 to 0,9 % of planning and construction costs, for smaller and/or very complex projects up to 2,0 %.

5.3 Project Control

When using project control, the client desires professional support in the areas of organization, schedule, costs and quality. He or she, however, wants to be in the managerial function himself or herself and intensively accompanies the entire project.

Just like project controlling, a project controller is on a staff function to the project management of the client. However, project control does not only supply control data but also works out organization, schedule and cost plans for project management and supports project management in an assisting function when it comes to coordination. Additionally, there are of course the same types of control functions as apply for project controlling.

Here, too, the project participants are in a managerial function to the client who assigns them directly. Depending on whether awarding is done to a general contractor or according to the individual trades, awarding of planning is undertaken either directly according to HOAI or by orienting itself on HOAI.

Contrary to its name, project control only indirectly exercises a control function via project management and only points out deviations from the target. However, this control function is greatly improved because target data are either determined by project control itself or are secured through undertaking control calculations.

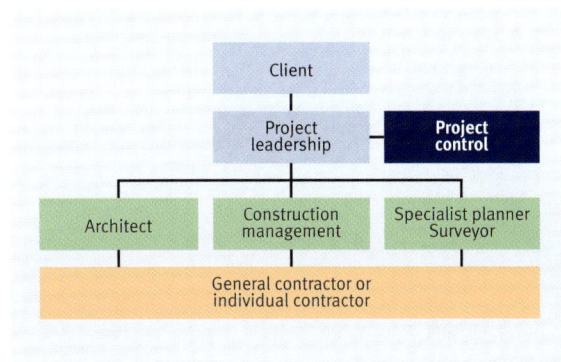

Fig. 5–3 Project control

This variety is used by professional clients with know-how of their own who, however, lack sufficient management capacity for special projects. As a rule, they assign project control to become the active support of their project management. Fees usually range in the area of ca. 1,1 to 1,5 % of planning and construction costs, for smaller and/or very involved projects also up to 2,4 %.

5.4 Project Management

In addition to the performances of project control, project management takes over the line function of the client toward the other project participants. As a rule, the client will assign the so-called delegable client functions. Often, this variety is used for more complex projects or by clients who do not have their own know-how nor their own management capacities.

Contrary to project control, the project manager is a kind of "Temporary CEO" acting on mandate by the client in managerial function with direct instruction authorization toward planners and executing entities. He/she influences proper project organization, executing strategy and processes, all the way to process optimization and process simulation (as an additional service). Another additional service is that he/she ensures impeccable and effective project communication and intensively consults the client when it comes to contract monitoring.

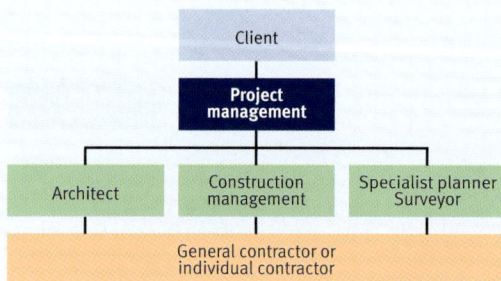

Fig. 5–4 Project management

Here, too, the other project participants are in managerial function to the client and are assigned directly by the same. Depending on whether awarding is done to a general contractor or according to the individual trades, awarding of planning is undertaken either directly according to HOAI or by orienting itself on HOAI.

Project management does have a direct control function, shows up target deviations and, by implementing suitable measures, ensures that the milestones set are actually also met. Target data are determined by project management itself and are additionally secured through sensitivity analyses. Good project management will generally try to make sure that the agreed cost and schedule targets are met down to the point and also ensure, largely, that quality targets are met. What is not included, however, is intensive optimization of planning contents themselves. This, in essence, is a matter for the architects and specialist planners.

This variety is used by clients without their own management know-how. They assign project management in order to then hand over project control to professional managers. Fees usually range around the region of ca. 1,6 to 2,1 % of planning and construction costs, for smaller or very complex projects also up to 3,0 %.

5.5 Construction Management (CM)

During construction management, the above-defined services of the project manager are built on in order to incorporate, into the project implementation, knowledge as early on as possible. Hence, additionally to the usual delegable client performance or service units, an original component is also, to a large part, being assigned. The client still only supplies the general guiding line, while the construction manager calls on his or her planning competency and management experience in order to transform this into a planning solution.

A construction manager, just like the project manager, stands in managerial function to the other project participants. He/she also has a right to contribute to decisions when it comes to selecting the project participants with

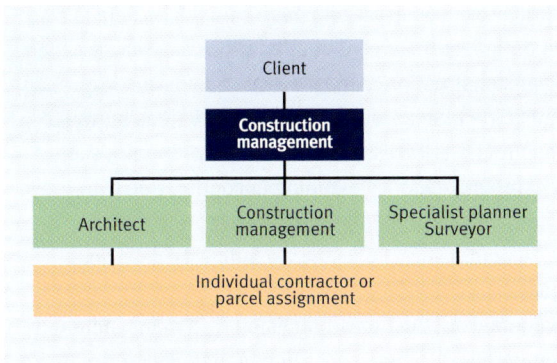

Fig. 5–5 Construction management (CM)

focus on processing procedure. All planning and implementation awards are assigned with involvement of and preparation by construction manager and directly entered into with the client. Due to the available management competency, the variety we call "construction management" uses individual planners and individual contractors or parcel assignments.

During the so-called "pre construction phase", the construction manager initially optimizes planning processes and planning content, based on his/her competency and experience. He/she does not furnish any planning services himself/herself according to HOAI, meaning that the often-mentioned "conflict of interest" cannot even result. By the planning process, already, a construction manager ensures that architectural and engineering solutions are able to be implemented in both a practical and also economical manner. Through using value engineering, the target definition of the project is repeatedly compared to the planning status and this procedure considers not only investment costs but also operating costs that will be incurred later on (Life cycle costs). Hence, professional plan coordination via planning management and content verification of plans is decisive for construction management.

During the "construction phase", the focus is on optimization of construction processes via "lean construction management", a construction site guide plan as well as on verification of feasibility via con-

struction process simulation. In order to keep targeted control of repetitive conflicts arising in planning and construction, there is an accompanying risk management. This renders transparent process risks as well as planning and construction risks and allows for them to either be eliminated or at least minimized. Anti claim management, which would need to be assigned separately, is either preventatively avoided by the presence of construction management or it can be professionally integrated.

This variety is used by clients without their own management know-how but who also expect intense and expert consulting and optimization of their projects from project management. Construction management hence, means a strong content component for them, which is why they expect only very experienced managers with planning and implementation experience. Accordingly, fees usually range in the area of ca. 2,2 % to 3,0 % of planning and construction costs, for smaller and/or very complex projects up to 3,9 %.

5.6   General Construction Management

In addition to the CM services, the GCM also provides planning services, especially conceptional planning services.

Increasingly, there are clients who only wish to have one contact person to accompany them through all the stages of performance. They do not see a conflict of interest in the so-called "One Stop Shopping" principle but, rather, have identified for themselves the advantages presented by clear responsibility definition and by avoiding interfaces. It is precisely these requirements that are met by general construction management through early incorporation of all specialist planners and, possibly, also of the architect. General construction management provides the complete master sectoral plan services, from concept to implementation planning, either by itself or through an outside partner acting on its mandate. This management variety requires a great deal of trust between client and construction manager. You have to know and respect each other!

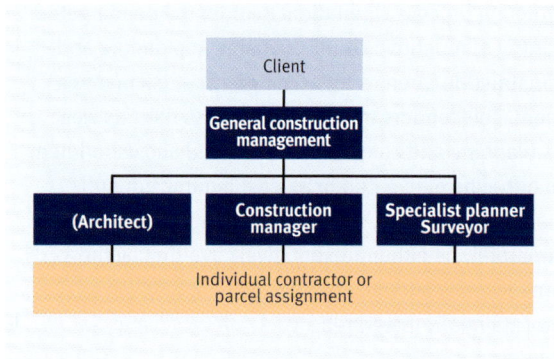

Fig. 5–6 General construction management

During the "construction phase", the focus here is also on optimization of construction processes via so-called "lean construction management", a construction site guide plan as well as monitoring of feasibility via construction process simulations. Additionally, there is complete logistics planning in cooperation with the construction companies and/or executing firms.

Object monitoring is undertaken by means of a "general construction supervision with management competency", where construction manager and project manager work together directly on location.

Reasons for assigning general construction management are principally the same as for construction management. However, the content component is even more magnified through direct acceptance of responsibility for planning, inclusive of implementation planning of the architect. This variety is preferably being applied for industrial construction, primarily also for "sustainable construction" as an alternative to a general planner, where, frequently, management competency and Value engineering come off badly.

Fees (excluding planning costs), due to the extra responsibility for planning, range somewhere around the area of ca. 3,1 % to 4,0 % of planning and construction costs, for smaller and/or very complex projects even up to 5,4 %. A flat fee offer including master section planning and project planning is submitted.

For this variety, it is often the case that, aside from the usual remuneration, additional premiums and bonus regulations and agreed upon, which are oriented either in total or in part on the degree of target achievement of the criteria agreed upon (schedule, costs, quality) and paid out accordingly. These bonus agreements ought, however, be regarded in a critical manner since they frequently place an extra stress on the required relationship of trust between client and construction manager (e.g. in the event of savings suggestions of any kind, which would increase the amount of the bonus but are not really desired by the client).

In relation to contract as well as in an organizational manner, a general construction manager holds a managerial function between client and planning participants. Here, architect may receive his/her assignment either directly through general construction management or through the client. In any event, general construction management needs to be granted instruction rights toward the architect if the task is to be optimally handled. Selection of the specialist planners is undertaken jointly with the client, but through the general construction manager.

A general construction manager also coordinates and optimizes planning services through directly working with the architect and the specialist planners while also involving the executing firms early on. This allows for incorporating their know-how, long before implementation commences, into the planning stages already. However, this requires an executive strategy that deviates from the usual HOAI stages and there also needs to be a step-by-step offer and awarding process where carefully selected and known firms initially are asked to submit bids on the basis of preliminary planning with requirement specification. Following planning optimization, the offers are then put into more concrete and detailed shape. Following intense inspection by general construction management, the awards are assigned in transparent agreement with the client.

5.7 Building Partner Management

During building partner management, general construction management is supplemented with select firms, the building partners, through takeover of the construction execution process. In contrast to a general contractor or main contractor solution, the individual companies are incorporated in a partner function into planning and preparation at a very early stage. The client thus receives the best possible cost and schedule security from medium-size enterprises that perform construction services in-house and are also available following commissioning.

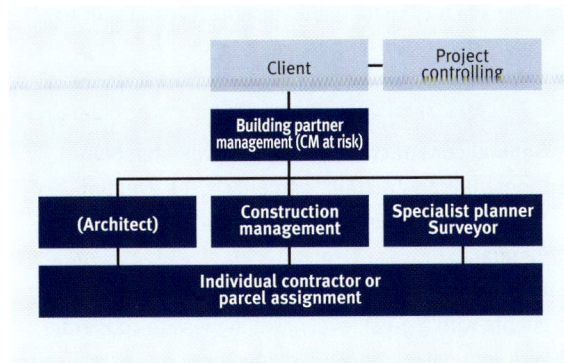

Fig. 5–7 Building partner management (CM at Risk)

In the event of large project management firms, their own companies are formed for building partner management and these firms specialize exclusively on this management variety. (At Drees & Sommer AG, for instance, this is the Building Agency.) Building partner management then incorporates the building partners into one project firm, where they are then liable to each other as well as in total to the client. Building partner management takes over the position of CEO for the project firm, which assigns any and all planning and construction tasks in its name. As a rule, about 80 % of task volume are handled by the partners of the project company and the rest is then assigned to outside contractors acting as sub-contractors.

The basic idea here is similar to that for construction management at risk but is still handled differently in the decisive details, for instance through the building partnership and through an extra segmenting of the so-called "pre-construction phase". As for general construction management, the executing firms are already integrated into the design phase and work out a fixed fee offer while adhering to the specified cost and schedule limitations. This offer must meet both quality and quantity specifications of the client. At this stage, the client can withdraw from the project if it turns out that his/her visions cannot be implemented. In this case, he or she is to pay the agreed-upon fee for services already rendered. If the flat fee offer is accepted, optimization of implementation planning can commence as it can for logistics and the processes for construction implementation. All this is done by means of an integrated planning process. Possible synergies are analyzed and interfaces between building partners are reduced and simplified. This way, for instance, construction logistics processes can already be coordinated early on. The client receives a total package, all one stop, and therefore can be sure that any arising conflicts are going to be resolved within the building partnership.

As a rule, this variety is only ever offered to solvent and trustworthy clients. Costs for management and risk coverage lie somewhere between 6,0 % and 8,0 % of planning and construction costs, for smaller and/or very complex projects up to 10 %. They are included in the flat fee offer.

The GMP variety (Guaranteed Maximum Price) has not been able to hold its own in general construction management, for similar reasons as the bonus approach. Clear fixed fee solutions are to be preferred for sure.

However, for this variety it is to be recommended that the client institutes his/or own project controlling that is independent of building partner management.

5.8 Comparison

All the varieties discussed have their advantages and disadvantages. The more outside know-how is purchased and the more responsibility assigned, the more there is a possible lever effect in favor of the client but the more, also, risk increases that the result does not live up to expectation if less competent or even shady partners are involved. Further, of course, the amount of fees to be paid also rises as delegated tasks and risk increase.

For the first of these four varieties we have just looked at (without HOAI and construction implementation services), there is the advantage that only delegable and original "client tasks" are assigned to an outside management firm.

Application areas are as follows:

– **Project controlling:** Very professional clients with their own know-how, e.g. a construction department of their own, or for very simple projects.
– **Project control:** Professional clients with know-how of their own but who lack management capacity for special projects. As a rule, they will assign project control for active support of their own project management.
– **Project management:** Clients who do not have management know-how of their own. These assign project management in order to hand over project supervision with professional managers.

– **Construction management:** Clients without their own management know-how but who additionally expect intensive and expert consulting and optimization for their project.

Fees are somewhere between 0,7 and just up to 4,0 %.

The last two varieties present the largest optimization potential from a content, economical and process-engineering point of view. However, they ought to only be entered into with highly professional and trustworthy (references!) partners. They are especially suitable in the event that a project is being supervised from the start and the focus is on integrated planning, as ought to be the case for "sustainable design" (Green Building).

Application areas are as follows:

– **General construction management:** Preferably industrial construction, primarily in the event of "sustainable design", as an alternative to a General Planner.
– **Building partner management:** As a rule: Solvent clients with a great level of willingness to cooperate, with whom a relationship of trust has been established already over many years.

Experience has shown that the fee scale for these varieties is somewhere between 3 and just up to 10 %, depending on project size and degree of difficulty.

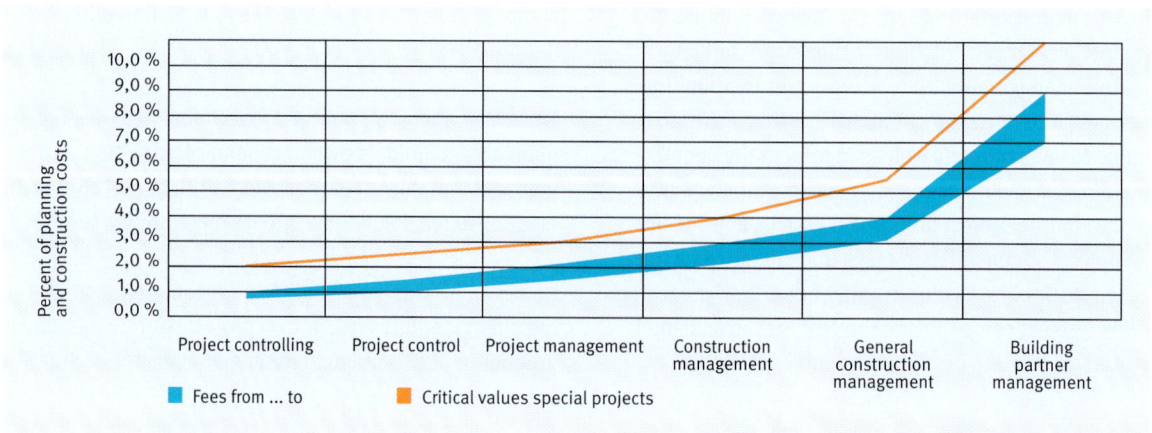

Fig. 5–8 Fee scale for the different management varieties

Index

Image Sources

© Jevgenijs Dasko – Fotolia.com (Cover)
Kroll Hamburg (P. 5)
© 2003 goebel photo (P. 5)
Hans-Georg Esch, Hennef (P. 5, 80)
Albrecht Haag Fotografie, Darmstadt (P. 5)
Martin J. Duckek, Ulm (P. 5, 88, 91)
Roland Halbe, Stuttgart (P. 5)
Brigida Gonzalez, Stuttgart (P. 5)
Vincent Mosch, Prof. Rudolf Schäfer (P. 5, 31, 49, 52, 60)
Tom Filipi, Stuttgart (P. 5)
Uwe Rau (P. 26)
Projektgesellschaft Neue Messe GmbH & Co. KG (P. 26)
Next Edit (P. 31)
XQ Media, Cologne (P. 31)
AOP Air shots Oliver Braitmaier, Stuttgart (P. 125, 127)